Roman A. Valiulin
NMR Multiplet Interpretation

Also of Interest

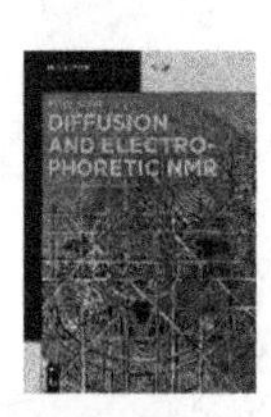

Diffusion and Electrophoretic NMR.
Stilbs, 2019
ISBN 978-3-11-055152-5, e-ISBN 978-3-11-055153-2

Compact NMR.
Blümich, Haber-Pohlmeier, Zia, 2014
ISBN 978-3-11-026628-3, e-ISBN 978-3-11-026671-9

Biomimetic Nanotechnology.
Senses and Movement
Mueller, 2017
ISBN 978-3-11-037914-3, e-ISBN 978-3-11-037916-7

Nanoparticles.
Jelinek, 2015
ISBN 978-3-11-033002-1, e-ISBN 978-3-11-033003-8

Roman A. Valiulin

NMR Multiplet Interpretation

An Infographic Walk-Through

DE GRUYTER

Author
Dr. Roman A. Valiulin

ISBN 978-3-11-060835-9
e-ISBN (PDF) 978-3-11-060840-3
e-ISBN (EPUB) 978-3-11-060846-5

Library of Congress Control Number: 2019937534

Bibliographic information published by the Deutsche Nationalbibliothek
The Deutsche Nationalbibliothek lists this publication in the Deutsche Nationalbibliografie; detailed bibliographic data are available on the Internet at http://dnb.dnb.de.

Typesetting: Integra Software Services Pvt. Ltd.
Printing and binding: CPI books GmbH, Leck
Cover image: Roman A. Valiulin (graphics), bestbrk / iStock / Getty Images Plus (background)

www.degruyter.com

The aim of art is to represent not the outward appearance of things, but their inward significance.

– Aristotle

Preface

The American Chemical Society (ACS) recognized the Varian A-60, among the first commercial Nuclear Magnetic Resonance (NMR) instruments for use in research, which was sold in the 1960s, as a U.S. National Historical Chemical Landmark.[1] Such applications of NMR were built on the foundational discovery and knowledge of many scientists who have also earned the Nobel Prize. In the decades since those seminal developments, NMR technology has provided immeasurable advantages for society, most saliently, as a diagnostic tool in the medical setting. It has also been critical as an analytical tool for chemists, providing an expedient and relatively straightforward means for determining molecular structure while preserving the sample. Specifically, for organic chemists, the interpretation of ^{1}H NMR multiplets is among the most useful techniques for determining the molecular structure and likely identity of numerous organic compounds.

More specifically, research chemists who intend to publish an ^{1}H NMR spectrum or characterize an organic compound are tasked with accurately and reliably reporting the H–H coupling constants to support their findings. However, the complexity of some ^{1}H multiplets often discourages the investment of time needed to fully document them. This may lead to either incorrect identification or gross simplification of ^{1}H NMR spectra descriptions, where numerous NMR signals are just described as a *multiplet* (**m**). When coupling constants are not extracted, useful structural information is lost. During the undergraduate and graduate years, ^{1}H NMR spectra interpretation can seem especially intimidating due to paucity of experience and practice. For example, it is not always intuitively clear how the complex *J*-tree diagrams can be built and what the values of the true coupling constants[2] should be. In these instances, elegant practical methods and intuitive mnemonics can aid the student and researcher to develop experience and to build confidence. Equipped with this approach, and incorporating my own knowledge and experience as a Ph.D. student, postdoctoral researcher, and as a teaching assistant and an individual tutor, I have effectively made this technique and skill become an integral part of my analytical toolkit. However, it took over a decade of study, practice, and experience to arrive at that point. While many excellent textbooks exist on the subject matter, there appears to be a general trend focusing on either basic examples or covering complex, rare, and even seemingly obscure ^{1}H NMR multiplet cases. Still other textbooks focus on the math and physics that undergird the technique, which, though essential, is not particularly accessible or useful for the organic chemist in their day-to-day research.

---

1 Accessed June 15, 2019 at https://www.acs.org/content/acs/en/education/whatischemistry/landmarks/mri.html

2 In this workbook we refer to the coupling constants that have physical meaning as "true" coupling constants. In the literature other terms may appear: "actual" or "real." The constants that are linear combinations of the true constants may be called "apparent," "not true," or "nonrelevant."

https://doi.org/10.1515/9783110608403-201

This workbook strives to fill that apparent gap in practice, by providing a focused, step-by-step, "walk-through" of ^{1}H NMR spectra interpretation. Moreover, it heavily capitalizes on our inherent nature as visual beings and visual learners, assembling the text around copious illustration, visualization, and infographic depictions. The book strives to be a practical visual guide (and can also be thought of as a "portable tutor") that is tailored for organic chemists. It aids the reader in the interpretation of the ^{1}H NMR signals: extraction of useful constants and properly naming the multiplets. It includes carefully sequenced, guided, practice problems and visually rich answer keys to help the reader develop familiarity and experience. It also presents salient trends and provides an overall conceptual framework for the reader to organize their understanding. Notably, this book is not intended to be a scientific paper nor a literature review, but is designed to be a visual pedagogical tool. Moreover, it omits discussion of NMR theory and should not be used as a comprehensive textbook that covers all aspects and nuances of NMR spectroscopy. Readers seeking to become more specialized in the subject-matter may wish to utilize this workbook in conjunction with dedicated coursework on spectroscopy. Where possible, references are made to relevant primary authority, should the reader seek a deeper understanding.

While modern advances in NMR processing software[3] have made manual ^{1}H multiplet decryption easier, mastering it is still a fundamental skill and essential for any researcher's analytic toolkit. Obtaining and reporting appropriate coupling constants remains just as useful for structure elucidation [1, 2] as choosing the appropriate pulse sequence (in two-dimensional NMR) [3, 4]. This book takes advantage of our inherent preference for visual comprehension, and will be even more useful to those who can quickly identify patterns, schemes, structures, and tendencies, by observation.

3 For example, *http://mestrelab.com/software/mnova/* (accessed June 15, 2019).

Contents

Foreword

I am a synthetic organic chemist by training and profession. I have also always been an amateur artist who has an eye for conveying information (or emotion) using color, depth, shape, space, and patterns, primarily through drawing. For many years, the realm of science and my hobby for art and drawing were two distinct worlds. When introduced to the concept of an "infographic", I immediately knew there was a place for these two worlds to meet. I started actively drawing infographic summaries, memory-aides, tables, flow charts, and other visuals in chemistry that I shared through my blog: ChemInfographic. To my pleasant surprise, there was a large audience of students and professionals with whom this mode of presentation resonated and who diligently followed my blog, actively engaging with it, and asking for more. In retrospect, it is not surprising, given that visuals can concisely convey dense amounts of information, when we thoughtfully use color, shape, space, and flow. Moreover, most people prefer visual learning and engagement, compared to other methods of communication. And that is where the inspiration for this book was born. I sought to think of the most useful analytical tool in my work as a synthetic organic chemist and challenged myself to develop a study-aid that was intuitive, easy to engage with, and had more visuals than text (my dream book). Moreover, I wanted to engage with the reader, to have a "portable tutor," where the reader could immediately absorb the information and "try their hand at it," or as others put it, I strived to create a "kinesthetic" learning tool.

– Roman A. Valiulin, Ph.D.

https://doi.org/10.1515/9783110608403-001

1 Introduction

This workbook provides a guided walk-through of ^{1}H NMR spectra interpretation, with a carefully sequenced collection of practical exercises designed to reinforce understanding of common concepts. The book focuses on *simple* and *complex* first-order ^{1}H NMR multiplets and is specifically designed for the needs of a developing organic chemist. It can be utilized as a companion to chemistry studies at the undergraduate as well as graduate level. Furthermore, experienced chemists seeking additional practice with the ^{1}H NMR interpretation can also utilize this book as a visual guide and a workbook.

As discussed before, the book does not present scientific findings and is not intended to be a comprehensive nor exhaustive study of the fundamental theory undergirding NMR. Moreover, those seeking to develop a specialization in this area should consider using this book in conjunction with a series of NMR spectroscopy courses for comprehensive coverage and knowledge.

This workbook embraces the intuitive and elegant algorithm previously presented by Professor Thomas R. Hoye and colleagues [5, 6], which enables a quick and very informative determination of coupling constant values. The literature may contain alternative ways for determining coupling constant values and students may wish to supplement their practice by exploring those publications, as they craft their own personal approach. Uniquely, as a student, teaching assistant, and graduate researcher not too long ago, the author is mindful of knowledge gaps that can present challenges for learning. Thus, the book strategically focuses on the most useful scenarios, provides an intuitive overview, presents a sequential and guided progression in the complexity of the problems, and highlights common trends to anticipate likely challenges that may arise in application. Most uniquely, it aims to provide a highly visual tutorial with diagrammatic and illustrative aids developed by the author. Readers who enjoy this highly visual pedagogical method are invited for further reading on the author's ChemInfographic Blog[4] and are invited to follow the author on Twitter™.[5]

The book can be divided into two parts. The first part introduces key concepts, defines main terms, describes a common nomenclature, and sets forth certain assumptions that are used throughout the workbook. Next, it provides detailed, step-by-step walk-throughs with multiple examples: three (3) total core examples, with five (5) steps covered in each example. Each step of every walk-through is supported by a rich illustration and practical tips in the form of mnemonic shortcuts and rules that are either commonly known in the field or suggested by the author. Readers may identify their own patterns and are encouraged to develop their own mnemonic devices and shortcuts to further reinforce learning. Each example concludes with a summary

4 Accessed June 15, 2019 at https://cheminfographic.wordpress.com/
5 @RomanValiulin

https://doi.org/10.1515/9783110608403-002

Infographic. Chapter 4 contains mnemonic rule summaries that classify some general trends and that help reinforce a conceptual framework of the entire process.

The second half of the book has a set of carefully sequenced exercises, followed by a set of graphical answers. Four (4) subsets are presented, progressing from simple to more complex: Elementary, Intermediate, Advanced, and Expert. Special attention is given to intermediate and advanced exercises because they represent the most common types of multiplets and are most likely to be encountered in application. The reader is encouraged to look over the entire workbook of problems and to develop a study plan that works best for their goal and level of understanding. Those for whom the entire subject-matter is new are encouraged to go in order and to complete all of the problems. More-experienced organic chemists may wish to skip ahead to the complex problems. All readers are encouraged to carefully follow the first half of the book and to take advantage of the illustrated and detailed answer keys for all of the exercises. Moreover, readers can find a few rich-color infographics at the end of the book (e.g., "Deuterated NMR Solvents") as useful quick reference guides.

2 Simple First-Order ^{1}H NMR Multiplets: Pascal's Triangle

The multiplets observed in ^{1}H NMR spectra can be grouped into three major categories: **First-Order** multiplets, **Second-Order** multiplets, and other **Higher-Order** multiplets (Figure 2.1).

Second-Order and **Higher-Order** multiplets are not covered in this workbook. For further reading please refer to more comprehensive NMR spectroscopy literature such as [7–9].[6]

First-Order multiplets are composed of two types: ***Simple*** and ***Complex*** *First-Order* multiplets (Figure 2.1). Without going into detailed discussion of the physics and theory of the Nuclear Magnetic Resonance spectroscopy, which is beyond the technical scope of this book, true ***Simple*** *First-Order* multiplets observed in ^{1}H NMR spectra obey certain rules:

- The number of peaks in the multiplet (multiplicity) can be determined from the **n + 1 rule**, where **n** is the number of equivalent neighboring H atoms.
- All the *J*-coupling constants to the neighboring protons are identical.
- The intensities (≈ heights) of the peaks in the multiplets can be derived from a mnemonic rule: using **Pascal's triangle** (Figure 2.2). For example, the intensities (≈ heights) of the peaks in a true *triplet* should be 1:2:1.
- The distances between each peak in a multiplet are identical and equal to the magnitude of the coupling constants measured in Hertz (Hz).

Additionally, we distinguish a subset of the *Simple First-Order* multiplets: ***Fundamental*** *Simple First-Order* multiplets (this is not an official term). These represent the simplest and most common examples of *Simple First-Order* multiplets and have a unique single letter notation: *singlet* (**s**), *doublet* (**d**), *triplet* (**t**), and *quartet* (**q**) (top four examples in Figures 2.2 and 2.3).

Complex *First-Order* multiplets, on the contrary, do not obey the *n+1 rule* and *Pascal's triangle* cannot be used to derive the peaks' intensities. A unique and defining feature of complex first-order multiplets is the fact that the *J*-coupling constants are not identical. In some instances, all the coupling constants to the neighboring protons are completely different: **dd**, **ddd**, **dddd**, **ddddd**, and so on. In other cases, some constants are identical and some are different: **dt**, **dq**, **dtd**, **qt**, and so on. For deeper knowledge on when a multiplet would be considered **First-Order** or **Second-Order**, refer to a few suggested NMR textbooks [7–9]; there are also several robust electronic

6 There are also numerous online tutorials and other resources, for example: https://www.chem.uci.edu/~jsnowick/groupweb/files/MultipletGuideV4.pdf or http://u-of-o-nmr-facility.blogspot.com/ (accessed June 15, 2019).

https://doi.org/10.1515/9783110608403-003

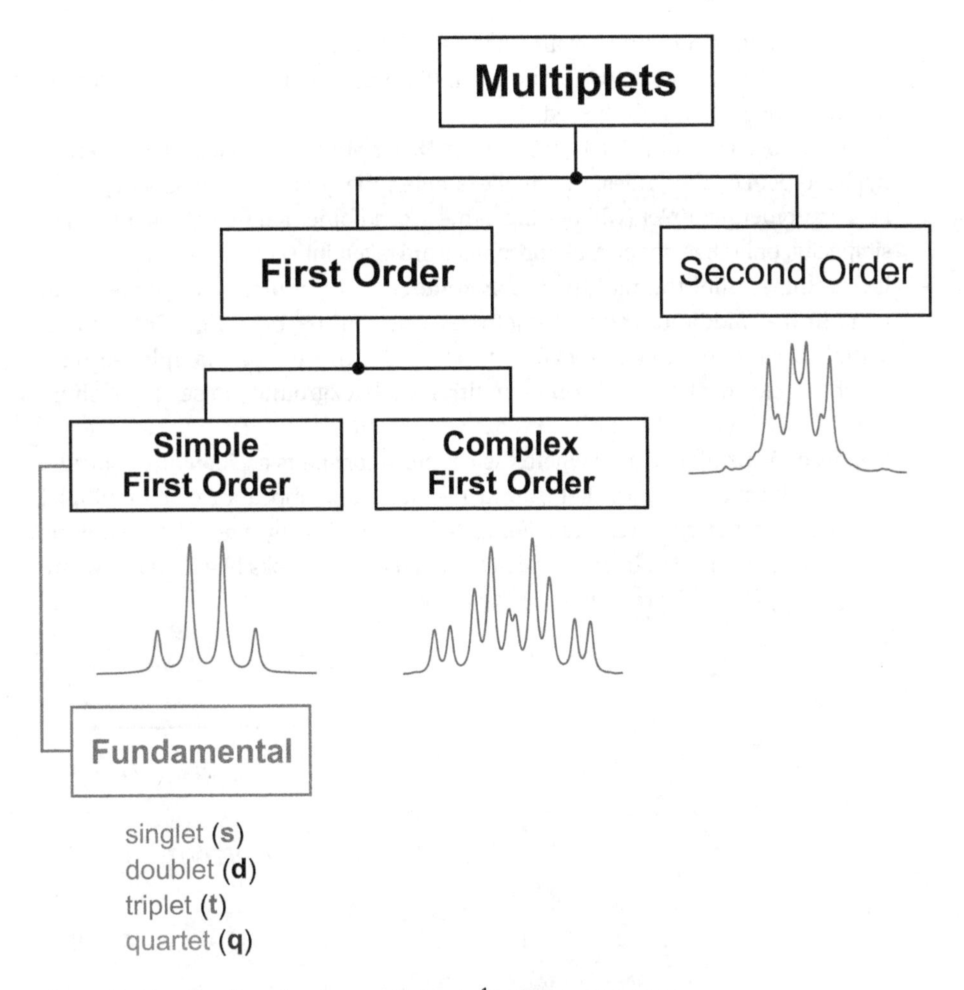

Figure 2.1: Classification of various multiplets in ¹H NMR spectroscopy.

resources: a particularly good example is an NMR spectroscopy course and manual by Professor Hans J. Reich [10].

Our tutorial in this book will focus only on the description of *Simple* and *Complex First-Order* multiplets. Before we start the visual walk-through, we should be cognizant of a few assumptions that have to be made while discussing the images of the multiplets.

- Only **¹H NMR** spectra are discussed in this tutorial.
- Each multiplet represents **1H** (single proton), that is, it should integrate as 1 (one) with respect to the other signals in the spectrum. Note, due to the symmetry or degeneracy, integration may be higher: 2H, 3H, . . ., 6H (e.g., two methyl groups in *iso*-propyl).

- Only **H–H coupling patterns** are considered, that is, each proton (represented by the multiplet) is coupled to one or multiple neighboring H atoms. Coupling to other nuclei are not discussed.
- Each multiplet is assumed to be a true **First-Order Multiplet**, that is, it is governed by first-order coupling rules. In actual examples, the multiplet may only *appear* to be a first-order multiplet (while being a true second-order multiplet by nature). For simplicity, only their shape and appearance are taken into consideration here.
- It is assumed that the multiplet is **symmetric**, with a plane of symmetry (**σ**) lying in the middle of the multiplet (i.e., it is centrosymmetric). Note that in actual cases there are obvious distortions and artifacts, for example, second-order effects, overlap with other multiplets, background noise, poor shimming, impurities, and so on.
- **Apparent Multiplets** (cases where the coupling constants accidentally coincide) are not subject to analysis, that is, only the shape and appearance of a multiplet recognized by the eye is reported. For instance, what is truly a *doublet of doublets* may appear as a *triplet*. If it is a *doublet of doublets* that looks like a *triplet* (where $J_1 \approx J_2$), a *triplet* will be reported with an assumption $J_1 = J_2$.

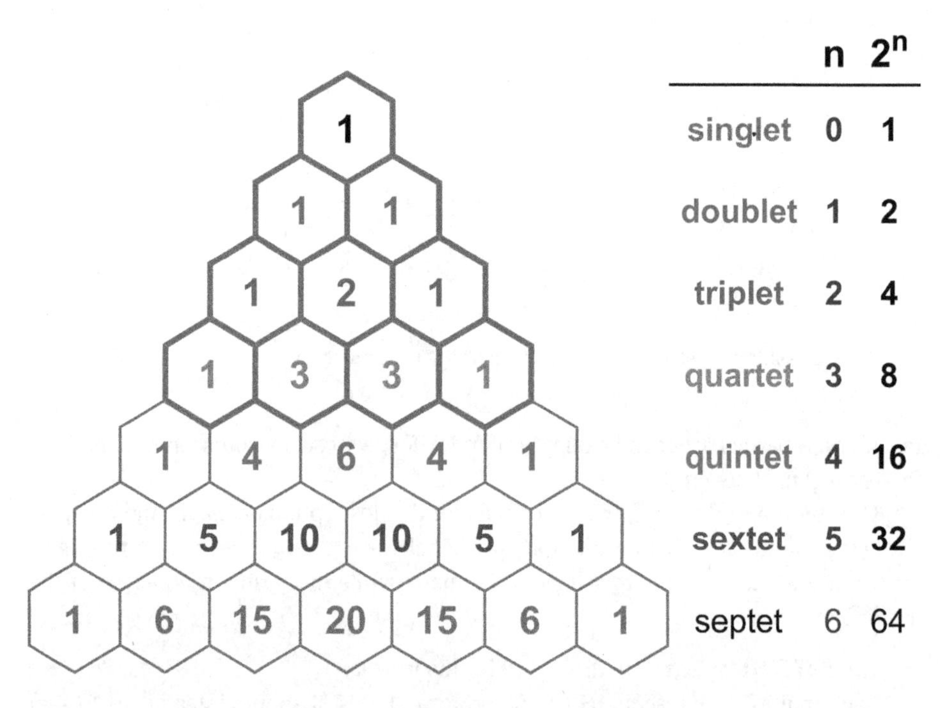

Figure 2.2: Pascal's triangle.

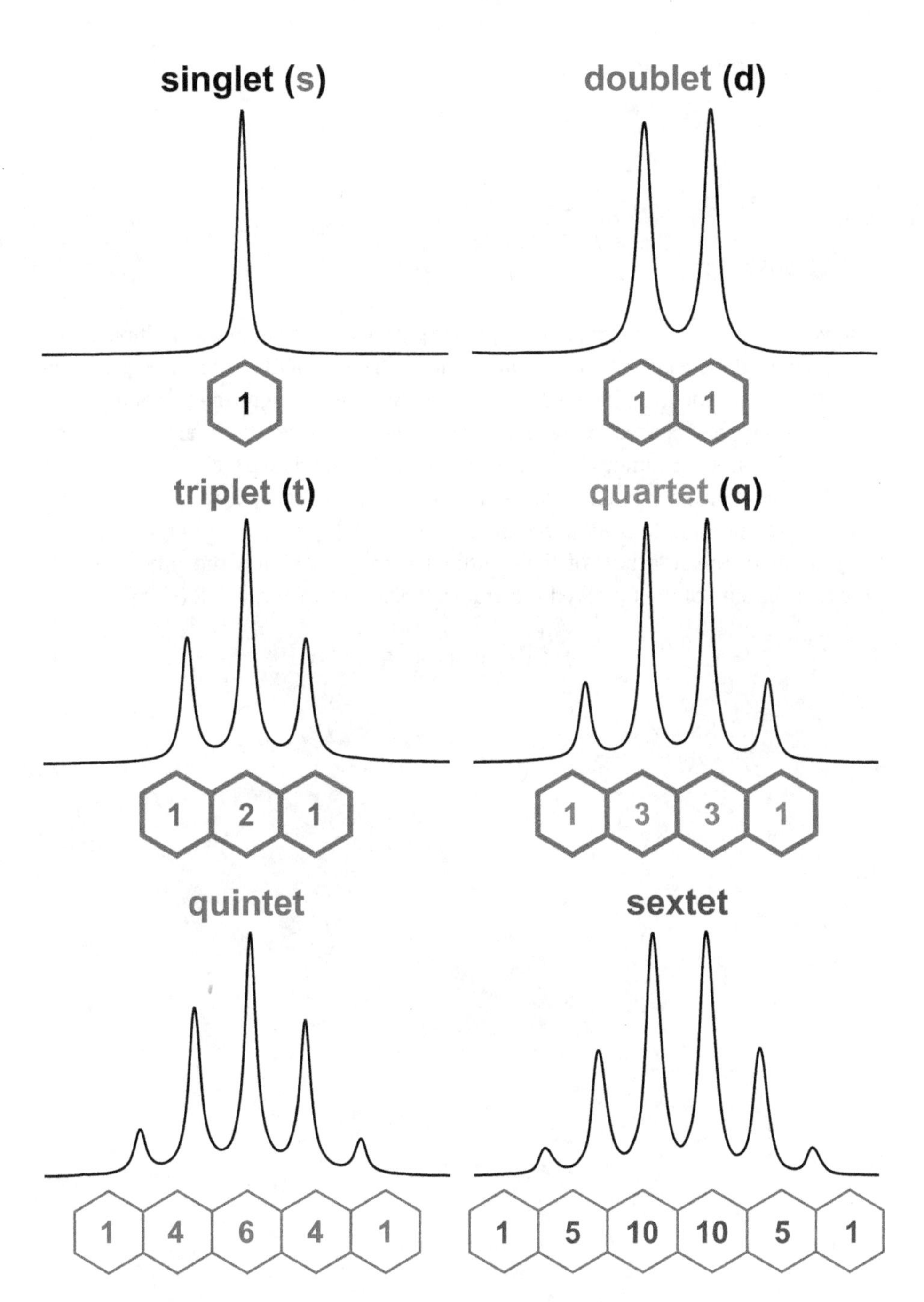

Figure 2.3: Simple First-Order multiplets, including the four Fundamental multiplets: **s, d, t, q**.

3 Complex First-Order ^{1}H NMR Multiplets: Infographic Walk-Through

3.1 Doublet of doublet of triplets [dd**t**]

3.1.1 Step 1: Integration

First, we determine the relative intensity (≈ height) of each peak in the multiplet. This can be simply done *qualitatively*: by comparing the heights of the peaks. This is not the most accurate method, but if the quality of the multiplet is high (the peaks are well-resolved, sharp, and spaced out without overlapping with other NMR multiplets or the peaks within the same multiplet), it is usually a quick and reliable technique.

Alternatively, the intensity can be measured *quantitatively*, by integrating the peaks using appropriate NMR processing software [11]. As shown in Figure 3.1, the integration[7] of the first half of the multiplet corresponds with the ratio 1:2:1:1:3. Note that the sum of the obtained integral values in this instance is **8** (eight):

$$1+2+1+1+3=\mathbf{8}$$

7 Each NMR signal that is discussed in this workbook is called a *multiplet*. Each multiplet is composed of several *peaks* of different (or similar) intensities. Here we use the terms "peak intensity" and "peak integration" interchangeably. In ideal cases, the peak intensity is proportional to "peak height," such that peak intensity = peak integration ≈ peak height.

https://doi.org/10.1515/9783110608403-004

STEP 1

1+2+1+1+3 = **8**

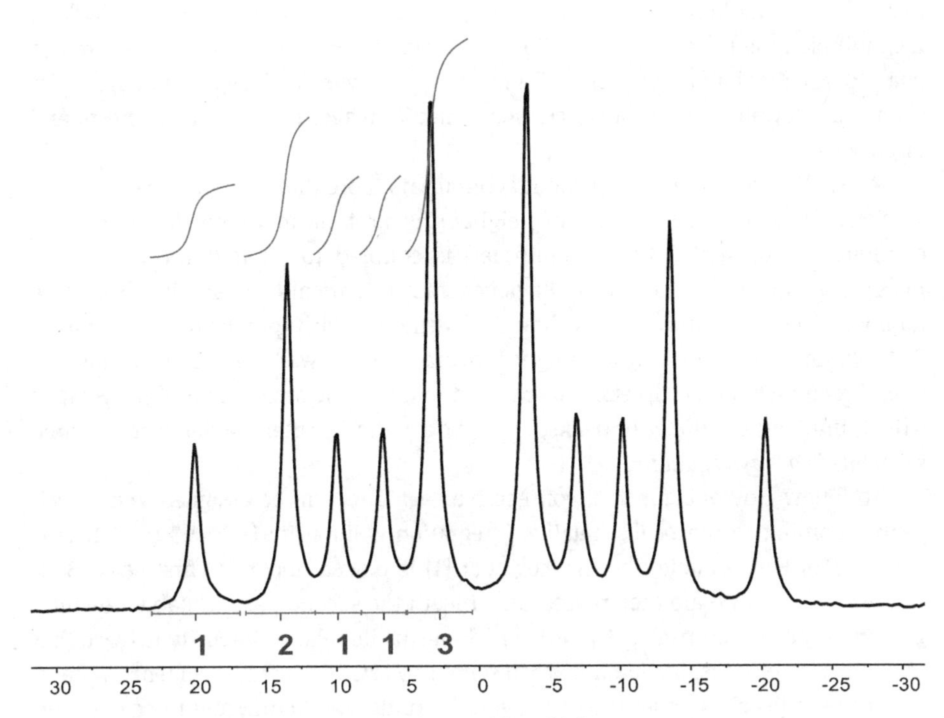

Figure 3.1: Integration of the individual peaks in a **ddt** multiplet.

3.1.2 Step 2: Symmetry and 2^n Rule

First-Order ^{1}H multiplets are centrosymmetric, and it is sufficient to analyze only the first half of the multiplet. Due to the *plane of symmetry* (σ) positioned in the middle of the multiplet, the remaining integration and assignment of the peaks will be a mirror image of the first half (i.e., 3:1:1:2:1). As mentioned previously, in actual examples, the shape can deviate from an ideal centrosymmetric image due to various effects and distortions.

All first-order ^{1}H multiplets should contain no more than $\mathbf{2^n}$ peaks of similar intensity: where $\boldsymbol{n}$ is the number of neighboring protons to which the H (proton) of interest (represented by the multiplet) is coupled to (or $\boldsymbol{n}$ is the number of *J*-coupling constants). Because of the degeneracy (identical *J*-values) the observable number of peaks is usually *less* than 2^n. In the scholarly publication by Thomas R. Hoye and colleagues [6] a straight-forward method was described to interpret *Complex First-Order* multiplets.[8] The method relies on the assignment of ***segments***,[9] which, unlike the number of peaks, always obey the $\mathbf{2^n}$ **rule:** *the full sum of all the segments is always equal to* 2^n.

To follow Hoye's algorithm, consecutive ***segments*** can be assigned under each peak according to the peaks' relative integration = intensity (1:2:1:1:3 [σ] 3:1:1:2:1) (Figure 3.2). For example, the first segment {1} is placed under the first peak. Both segments {2} and {3} are then placed underneath the second peak. Notably, the integration of the second peak is 2 (two), which means its relative intensity is twice that of the first peak; thus, it should be represented by two degenerate segments: {2} and {3}. Segment {4} goes under the third peak; {5} under the fourth; and three degenerate segments {6}, {7}, and {8} under the fifth, since the integration of the fifth peak is 3 (three), that is, it is three times taller relative to the first peak. This exercise extrapolates peak intensities (1:2:1:1:3) into segments {1}, {2+3}, {4}, {5}, {6+7+8} placed under each peak, according to their intensity (or integration):

$$1{:}2{:}1{:}1{:}3 \rightarrow \{1\}, \{2+3\}, \{4\}, \{5\}, \{6+7+8\}$$

The remaining assignment of the multiplet is a mirror image of the first half:

$$3{:}1{:}1{:}2{:}1 \rightarrow \{\mathbf{9}+\mathbf{10}+\mathbf{11}\}, \{\mathbf{12}\}, \{\mathbf{13}\}, \{\mathbf{14}+\mathbf{15}\}, \{\mathbf{16}\}$$

The numerical value (number) of the last assigned segment in this case is {**16**} (sixteen). It is also equivalent to the sum of all integral values (peak intensities in Figure 3.1):

8 There are other complementary algorithms derived to analyze first-order coupling patterns [12].
9 In the scholarly publication, the authors call them "units of intensity" or "components."

$$(1+2+1+1+3)+(3+1+1+2+1)=\mathbf{8}+\mathbf{8}=\mathbf{16}$$

As mentioned earlier, this value should always be equal to $\mathbf{2^n}$, where $\boldsymbol{n}$ = **4** in this specific case: $\mathbf{2^4}$ = **16.**

Mnemonic Rule I: *(a) For every first-order* ^1H *multiplet, the sum of the integral values of each peak within the multiplet should be equal to the numerical value of the last assigned* **segment***. (b) Note that according to the* $\mathbf{2^n}$ ***rule****, the full sum of the segments always equals* $\mathbf{2^n}$, $\boldsymbol{n}$ *is the number of J-coupling constants (N_J) or the number of neighboring protons to which the H of interest, represented by the multiplet, is coupled to.*

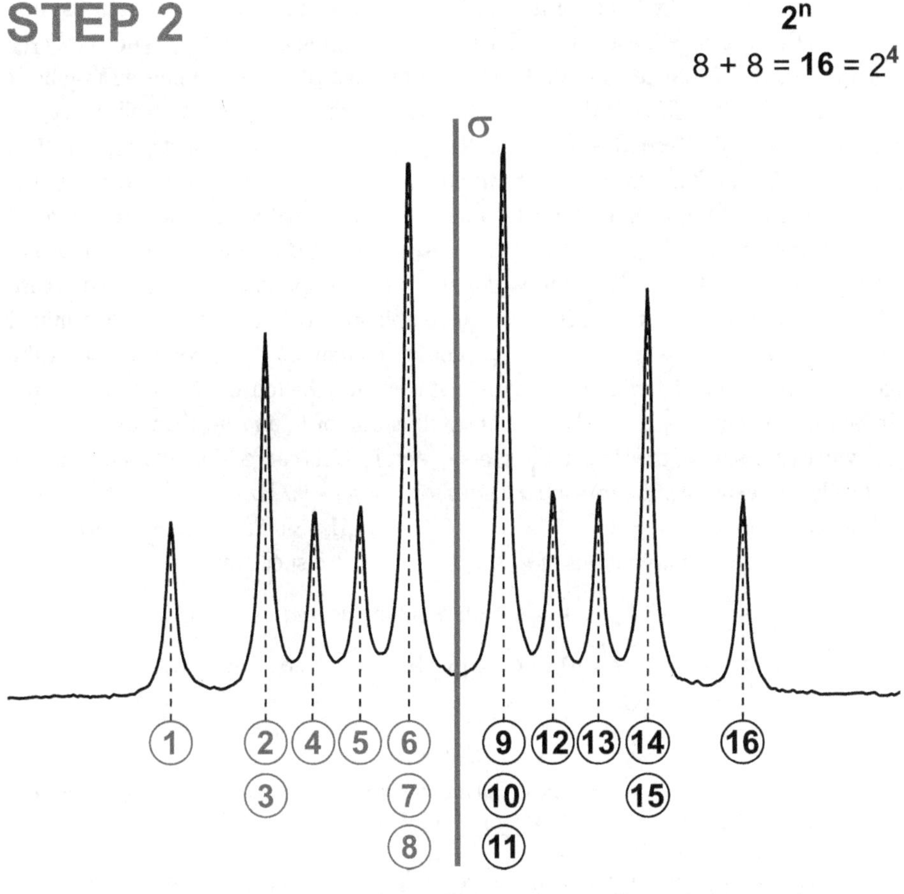

Figure 3.2: Numbering the individual peaks in a **ddt** multiplet according to their integration.

3.1.3 Step 3: Coupling Constants

Having determined that the sum of integral values is 16 (sixteen) and applying **Mnemonic Rule I** retrospectively, we can now determine that there are 4 (four) true *J*-coupling constants: if $16 = 2^n$, then $\log_2 16 = \mathbf{4}$ and $\boldsymbol{n} = \mathbf{4}$.

Therefore, we can describe this multiplet as a *doublet of doublet of doublet of doublets*[10] (**dddd**), in the most general terms. We can use this information as a guide during *J*-coupling constant extraction. For *Simple First-Order* multiplets, the distances between each peak in a multiplet are identical and equal to the coupling constant measured in Hertz (Hz). It is important to remember that for *Complex First-Order* multiplets the distances between the peaks are not identical. Moreover, not every measured distance will be proportional to a true coupling constant. Determining the correct value for each coupling constant and distinguishing true coupling constants from their linear combinations is one of the most challenging tasks.

Starting from the first peak on the left, labeled {1}, the smallest true coupling constant can always be identified for any multiplet, that is, the distance between the first peak and the second will equal the value of a true coupling constant ($\boldsymbol{J_{1\to2}}$ or simply $\boldsymbol{J_{12}}$) (Figure 3.3). This distance can also be represented as a blue vector between number 1 and 2 (1⟶2).[11] We will use this vector notation for other examples as well. Since the integral value of the second peak is two, there should be another constant $\boldsymbol{J_{1\to3}}$ (= $\boldsymbol{J_{13}}$), identical to J_{12}. Applying the same algorithm, we can measure the distance between the first and the third peak (blue vector 1⟶4). This vector's length is unique; thus, it must be the third true coupling constant $\boldsymbol{J_{1\to4}}$ (= $\boldsymbol{J_{14}}$). It is important to remember that not every measured distance will be proportional to a true coupling constant (unlike in case of *simple first-order* multiplets). Other linear combinations of the previously determined true coupling constants ($\boldsymbol{J_{12}}$, $\boldsymbol{J_{13}}$, and $\boldsymbol{J_{14}}$) should be eliminated. A good example is the red vector 1⟶5. The distance between the first peak and the fourth (1⟶5) is not unique. Apparent constant $\boldsymbol{J_{1\to5}}$ is merely a linear combination of J_{12} and J_{13}; thus, the length of $J_{1\to5}$ will be the sum of the length of vectors J_{12} and J_{13}. This can be demonstrated mathematically. For example, the *measured* values for $J_{12} = J_{13} = 6.7$ Hz and $J_{1\to5} = 13.3$ Hz. The *calculated* value for $\boldsymbol{J_{1\to5}} = J_{12} + J_{13} = 6.7 + 6.7 = 13.4$ Hz. Similarly, the distances 1⟶6 and 1⟶7 are linear combinations of $J_{14} + J_{12}$ and $J_{14} + J_{13}$, respectively:

$$\boldsymbol{J_{1\to6}} = J_{14} + J_{12} = 10.1 + 6.7 = 16.8\,\text{Hz}\ (\textit{measured as 16.8 Hz})$$

$$\boldsymbol{J_{1\to7}} = J_{14} + J_{13} = 10.1 + 6.7 = 16.8\,\text{Hz}\ (\textit{measured as 16.8 Hz})$$

10 Grammatically, it is more appropriate to call **dddd**: a *doublet of* ***doublets*** *of* ***doublets*** *of doublets*. However, in the NMR literature the plural form of the middle multiplets is often omitted: a *doublet of doublet of doublet of doublets*.

11 From here on, we will represent only true coupling constants using $\boldsymbol{J_{1n}}$ notation. Their linear combinations and any nonrelevant vectors will be depicted as $\boldsymbol{J_{1\to n}}$.

STEP 3

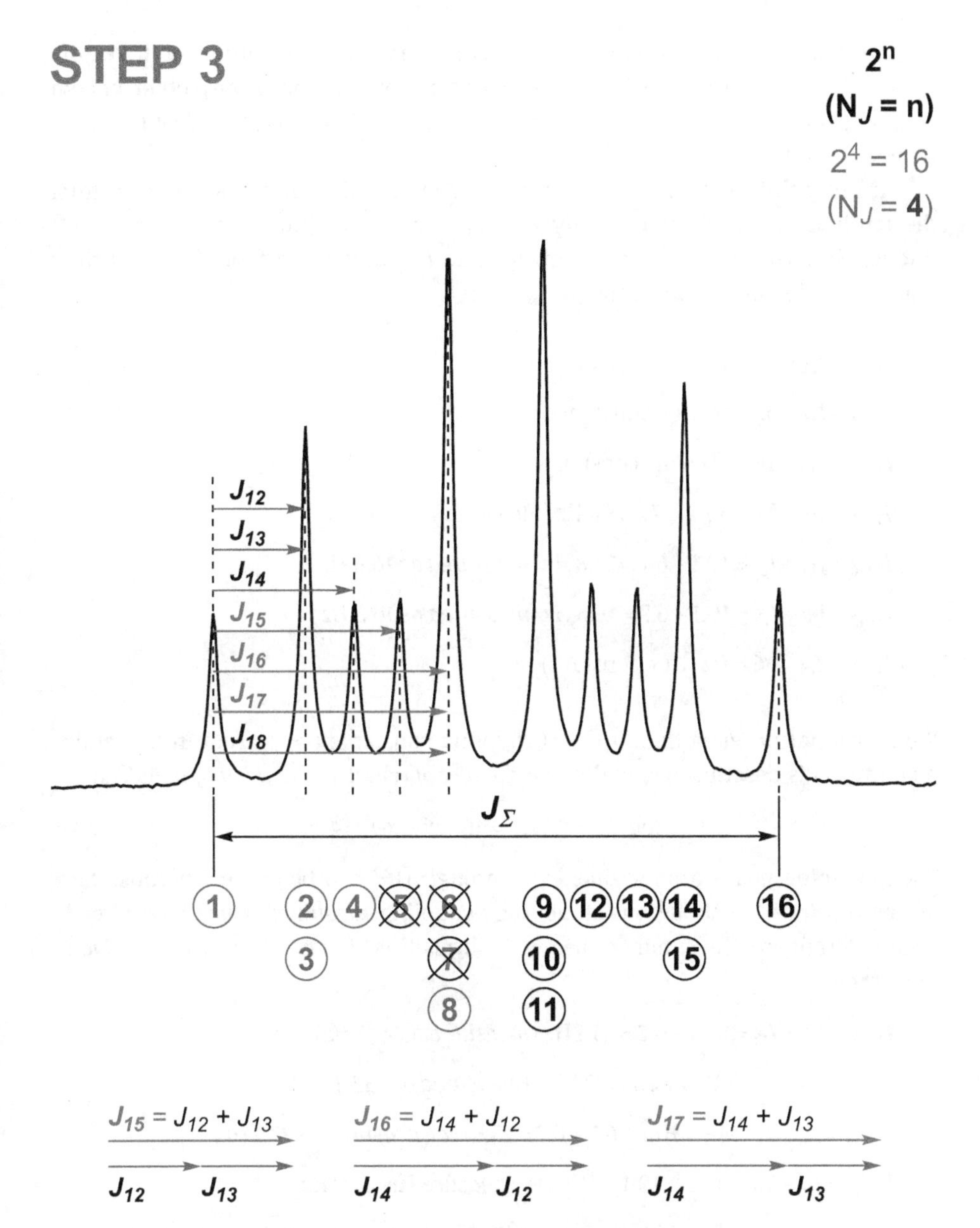

Figure 3.3: Identification of the *J*-coupling constants in a **ddt** multiplet.

The remaining blue vector 1→8 ($J_{1\to8}$) is identical by length to both vectors 1→6 and 1→7; however, it cannot be a linear combination of any other known coupling constants at this point. This means that constant J_{18} is the last true coupling constant.

At this point, it is wise to summarize the information that was extracted from the multiplet analysis: particularly first half of the multiplet from segment {1} through {8}. We identified the following true *J*-coupling constants (J_{1n}) and their nonrelevant linear combinations ($J_{1\to n}$):

$$J_{1\to2} = J_{12} = 6.7 \text{ Hz (true constant)}$$

$$J_{1\to3} = J_{13} = 6.7 \text{ Hz (true constant)}$$

$$J_{1\to4} = J_{14} = 10.1 \text{ Hz (true constant)}$$

$$J_{1\to5} = J_{12} + J_{13} = 6.7 + 6.7 = 13.4 \text{ Hz } (\textit{measured as 13.3 Hz})$$

$$J_{1\to6} = J_{14} + J_{12} = 10.1 + 6.7 = 16.8 \text{ Hz } (\textit{measured as 16.8 Hz})$$

$$J_{1\to7} = J_{14} + J_{13} = 10.1 + 6.7 = 16.8 \text{ Hz } (\textit{measured as 16.8 Hz})$$

$$J_{1\to8} = J_{18} = 16.8 \text{ Hz (true constant)}$$

We know that no more than 4 (four) coupling constants are present in this multiplet; thus, it is described generally as a *doublet of doublet of doublet of doublets*:

$$\log_2 16 = 4 \text{ or } 2^n = 16, \text{ where } n = 4$$

The remaining peaks from segment {9} through {16} can be omitted because they are simply linear combinations of the true *J*-coupling constants, which have already been determined. The summary below confirms this statement (*measured value in parentheses*):

$$J_{1\to9} = J_{18} + J_{12} = 16.8 + 6.7 = 23.5 \text{ Hz } (\textit{measured as 23.5 Hz})$$

$$J_{1\to10} = J_{18} + J_{13} = 16.8 + 6.7 = 23.5 \text{ Hz } (\textit{measured as 23.5 Hz})$$

$$J_{1\to11} = J_{14} + J_{13} + J_{12} = 10.1 + 6.7 + 6.7 = 23.5 \text{ Hz } (\textit{measured as 23.5 Hz})$$

$$J_{1\to12} = J_{18} + J_{14} = 16.8 + 10.1 = 26.9 \text{ Hz } (\textit{measured as 27.0 Hz})$$

$$J_{1\to13} = J_{18} + J_{13} + J_{12} = 16.8 + 6.7 + 6.7 = 30.2 \text{ Hz } (\textit{measured as 30.2 Hz})$$

$$J_{1\to14} = J_{18} + J_{14} + J_{12} = 16.8 + 10.1 + 6.7 = 33.6 \text{ Hz } (\textit{measured as 33.7 Hz})$$

$$J_{1\to15} = J_{18} + J_{14} + J_{13} = 16.8 + 10.1 + 6.7 = 33.6 \text{ Hz } (\textit{measured as 33.7 Hz})$$

The final distance from {1} to {16}, which can be represented by the longest vector $J_{1\to16}$, is a liner combination of all four (4) true coupling constants (J_{12}, J_{13}, J_{14}, and J_{18}):

$$J_{1\to16} = J_{18} + J_{14} + J_{13} + J_{12} = 16.8 + 10.1 + 6.7 + 6.7 = 40.3\,\text{Hz}\ (\textit{measured as } 40.4\,\textit{Hz})$$[12]

Going through the analysis described earlier, an interesting observation can be made, which is summarized as follows.

Mnemonic Rule II: *The distance between the very first peak {1} in a first-order ^1H multiplet and the second peak {2} (depicted as 1→2) will always equal the value of the smallest true coupling constant in any multiplet ($J_{1\to2} = J_{12}$).*

12 Please note that this example is elaborated further in **step 4**.

3.1.4 Step 4: *J*-Tree Diagram

Extracting all of the coupling constants from the multiplet significantly facilitates the prediction and reconstruction of the *J*-tree diagram (if needed). Often, drawing such diagram is not necessary, especially if the values of the true coupling constants were already identified. However, for educational purposes, it is a valuable exercise. The *J*-tree diagram is also a very informative visual tool that can help to connect the pieces of the conceptual "puzzle." It shows how the multiplet assumed its unique shape and how each peak gained its relative intensity. The *J*-tree diagram can be reconstructed easily by starting from the middle of the multiplet and drawing the first two branches of the tree separated by the longest vector (Figure 3.4), which represents the largest identified true coupling constant (blue vector 1→8 represented by $\boldsymbol{J_{18}}$). Each branch itself is split again proportionally to the second largest constant ($\boldsymbol{J_{14}}$), and so on, until all of the 4 (four) constants are accounted for ($\boldsymbol{J_{18}}$, $\boldsymbol{J_{14}}$, $\boldsymbol{J_{13}}$, $\boldsymbol{J_{12}}$). A more detailed, five-step procedure for the generation of a tree diagram can be found in Reference [5]; alternative methods also exist, for example [12]. Interestingly, the addition of all the blue vectors together (or the corresponding values of the true *J*-coupling constants measured in Hz) will equal the measured distance between the frontier peaks of the multiplet. In other words, the *measured* distance from segment {1} to {**16**} and represented by vector $\boldsymbol{J_{\Sigma}}$ is equal to the *calculated* value, represented by $\boldsymbol{J_{1\to16}}$:

$$\boldsymbol{J_{1\to16}} = \boldsymbol{J_{18}} + \boldsymbol{J_{14}} + \boldsymbol{J_{13}} + \boldsymbol{J_{12}} = 16.8 + 10.1 + 6.7 + 6.7 = 40.3\,\text{Hz}$$

$$\boldsymbol{J_{\Sigma}} = 40.4\,\text{Hz}\ (\textit{measured})$$

This rule of thumb is helpful when one of the coupling constants cannot be determined easily due to overlapping signals, for instance. There may be examples when the unknown constant $\boldsymbol{J_X}$ can be easily calculated by subtracting the rest of the true known constants from the measured $\boldsymbol{J_{\Sigma}}$ value.

Mnemonic Rule III: *The distance between the very first peak in a first-order ^{1}H multiplet {1} and the very last one {**16**} will always equal the sum of all true coupling constants in the multiplet ($\boldsymbol{J_{1\to16}} = \boldsymbol{J_{18}} + \boldsymbol{J_{14}} + \boldsymbol{J_{13}} + \boldsymbol{J_{12}} = \boldsymbol{J_{\Sigma}}$).*

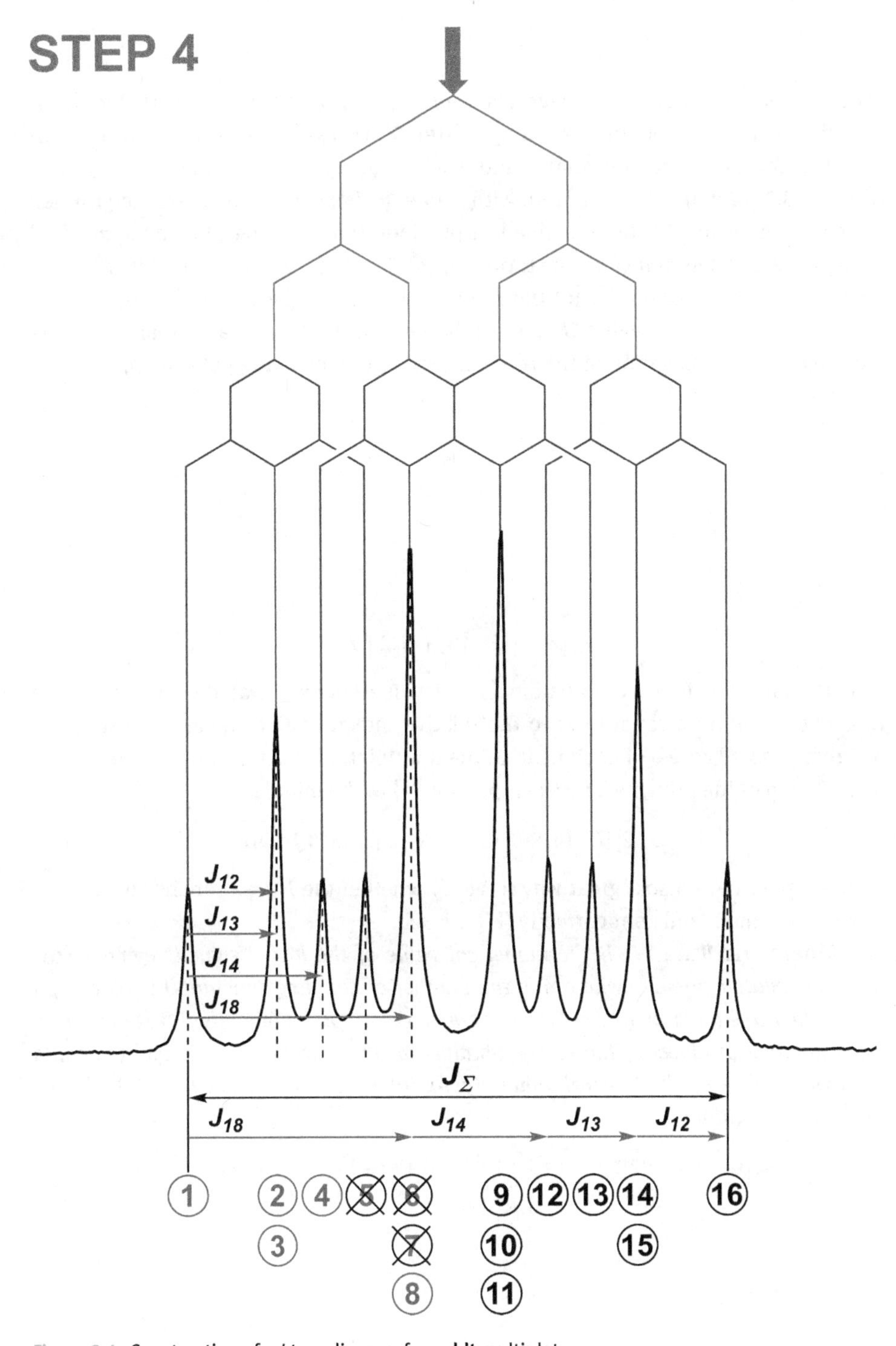

Figure 3.4: Construction of a J-tree diagram for a **ddt** multiplet.

3.1.5 Step 5: The completed “Puzzle”

The author finds the *J*-tree diagrams very helpful, albeit unnecessary for the *J*-coupling determination. In this example (**step 5**), we can demonstrate how the multiplet assumes its unique shape and each peak gains its intensity (≈ height) (Figure 3.5). For this, let us start with value **16** (sixteen), which we determined using the mnemonic rules described in previous steps: through (a) the sum of all integral values determined from **step 1**; (b) the 2^n rule ($2^4 = 16$), where n is the number of *J*-coupling constants; (c) the numerical value of the last assigned segment {**16**}, and so on. If this value **16** is placed on top of the *J*-tree diagram and split according to the propagation of the *J*-tree branches, the following pattern appears:

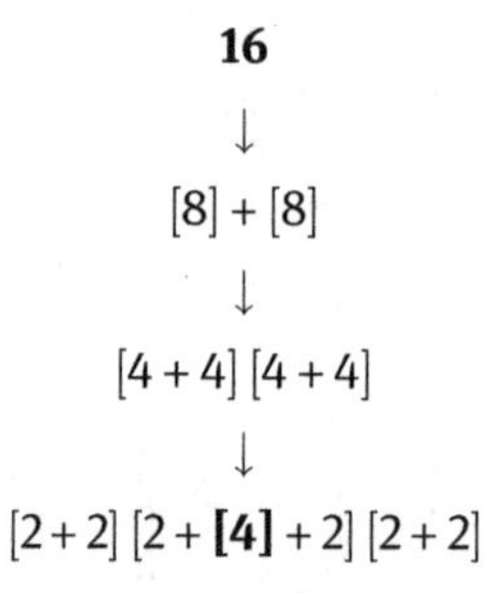

Additionally, the final combination of the values obtained at the bottom of the branches should be equal (⟷) to the relative integral values (peak intensity or ≈ heights in some cases) of each peak in the multiplet. Please take into consideration the overlap of the branches, for example, for **[4]** in the middle:

$$[2+2]\ [2+\mathbf{[4]}+2]\ [2+2] \rightarrow 1|2|1|1|3||\mathbf{3}|\mathbf{1}|\mathbf{1}|\mathbf{2}|\mathbf{1}$$

This simple rule is also a great way to verify whether the *J*-tree branches were positioned, assigned, and split correctly (Figure 3.5).

Mnemonic Rule IV: *If the numerical value of the last assigned segment {16} (also calculated from 2^4, where 4 is the number of coupling constants) is placed on top of the J-tree diagram and split according to the propagation of the J-tree branches, then the final products of the values obtained at the bottom of the branches should always equal (⟷) the integral values (peak intensities or ≈ peak heights) of each peak in the multiplet:*

$$1|2|1|1|3||\mathbf{3}|\mathbf{1}|\mathbf{1}|\mathbf{2}|\mathbf{1} \leftrightarrow (1+2+1+1+3)+(3+1+1+2+1) = 8+8 = \mathbf{16}$$

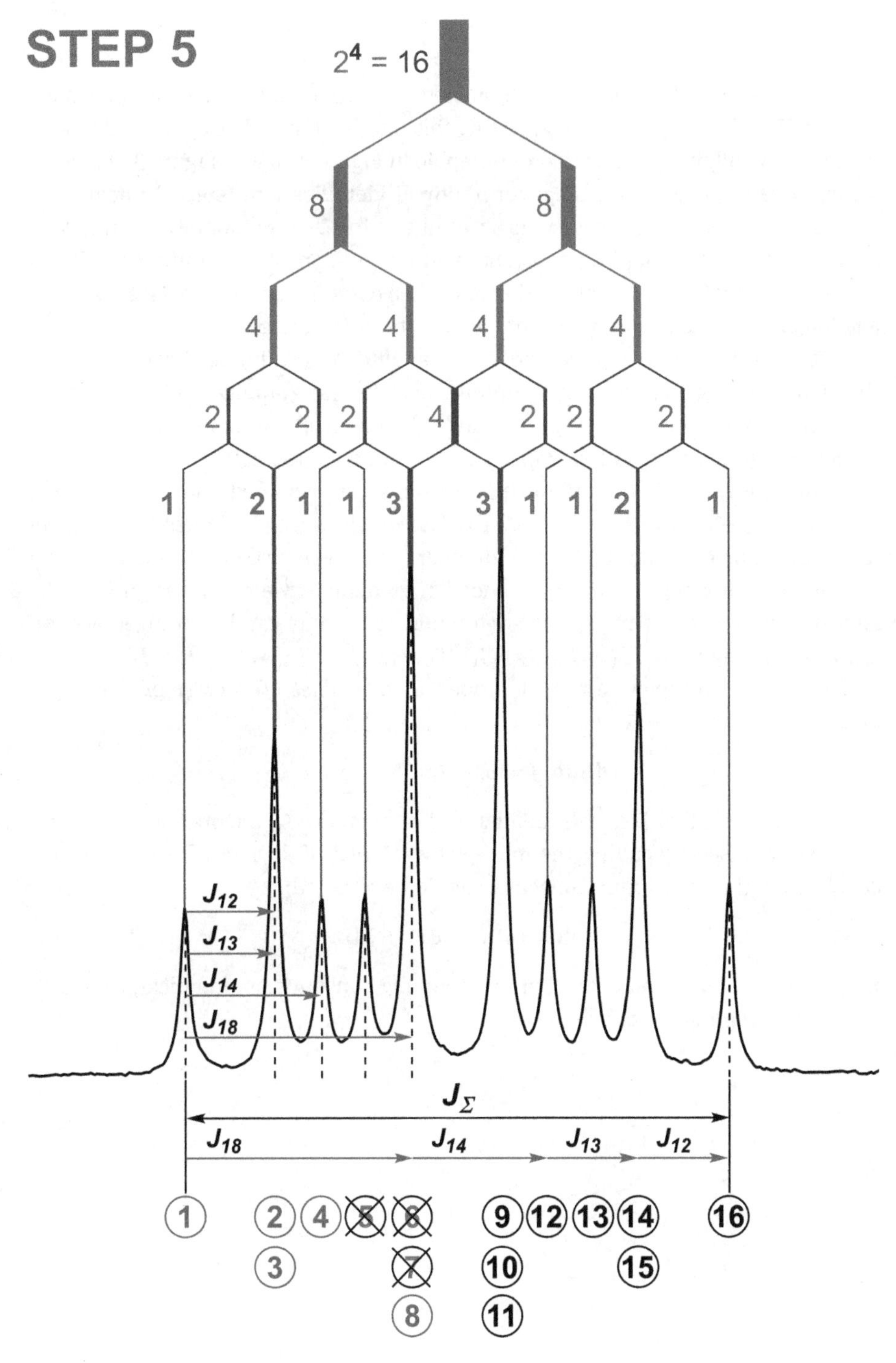

Figure 3.5: Summary illustrating the analysis of a **ddt** multiplet.

3.1.6 The Infographic [ddt]

Figure 3.6 provides an opportunity to immediately apply the five-step approach detailed earlier. Please practice interpreting this multiplet and check your work in the immediately following illustrated infographic in Figure 3.6a and Figure 3.6b, which summarizes the five (5) key steps for multiplet identification: (**step 1**) integration; (**step 2**) assignment of the segments according to the 2^n rule; (**step 3**) identification of the relevant true coupling constants and elimination of the nonrelevant linear combinations; (**step 4**) reconstruction of a *J*-tree diagram; and finally (**step 5**) propagation of the *J*-tree diagram and completion of the assignment.

Figure 3.6b also illustrates **step 3** with additional detail. Specifically, it shows the elimination of the apparent coupling constants (a) graphically (by comparing the vector lengths, e.g., $J_{1\to6} = J_{14} + J_{12}$) and (b) numerically by comparing constants' values measured in Hz. For example, the measured value for $J_{1\to6}$ = 16.8 Hz. Since the values for J_{14} = 10.1 Hz and for J_{12} = 6.7 Hz, their calculated linear combination will equal: $J_{14} + J_{12}$ = 10.1 + 6.7 = 16.8 Hz. This quick check eliminates $J_{1\to6}$ as a potential coupling constant, together with other nonrelevant vectors: $J_{1\to5}$ and $J_{1\to7}$.

Finally, following prevailing publication guidelines, we can reassign $\boldsymbol{J_1}$ as the largest coupling constant (J_{18}). The determined constants can be arranged according to their magnitude: $\boldsymbol{J_1} = J_{18}$, $\boldsymbol{J_2} = J_{14}$, $\boldsymbol{J_3} = J_{13}$, $\boldsymbol{J_4} = J_{12}$, where $\boldsymbol{J_1} > \boldsymbol{J_2} > J_3 = J_4$. Generally, the multiplet can be described as a *doublet of doublet of doublet of doublets*:

dddd, *J* = 16.8, 10.1, 6.7, 6.7 Hz

To reflect the fact that $\boldsymbol{J_3}$ and $\boldsymbol{J_4}$ are identical, that is, $J_3 = J_4$, a more specific assignment can be used by calling the multiplet a *doublet of doublet of triplets*. In this case, only 3 (three) coupling constants should be reported:

ddt, *J* = 16.8, 10.1, 6.7 Hz

Here, the term *triplet* indicates that two coupling constants are identical; thus, it is sufficient to report only one value (6.7 Hz).

I.

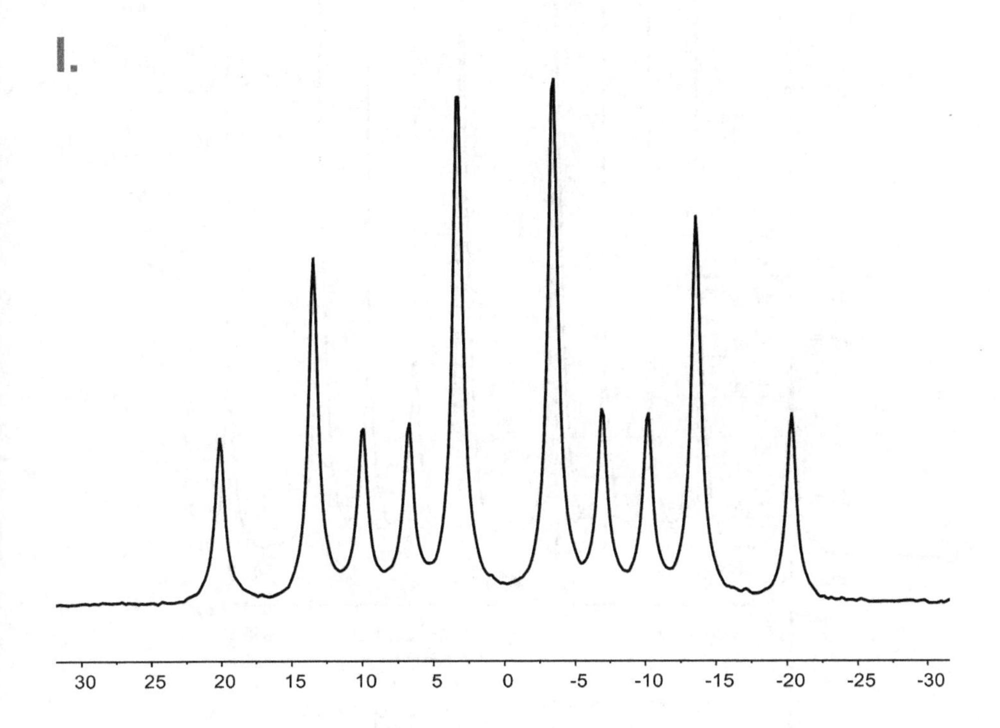

3.6: Advanced exercise **I** [**ddt** multiplet].

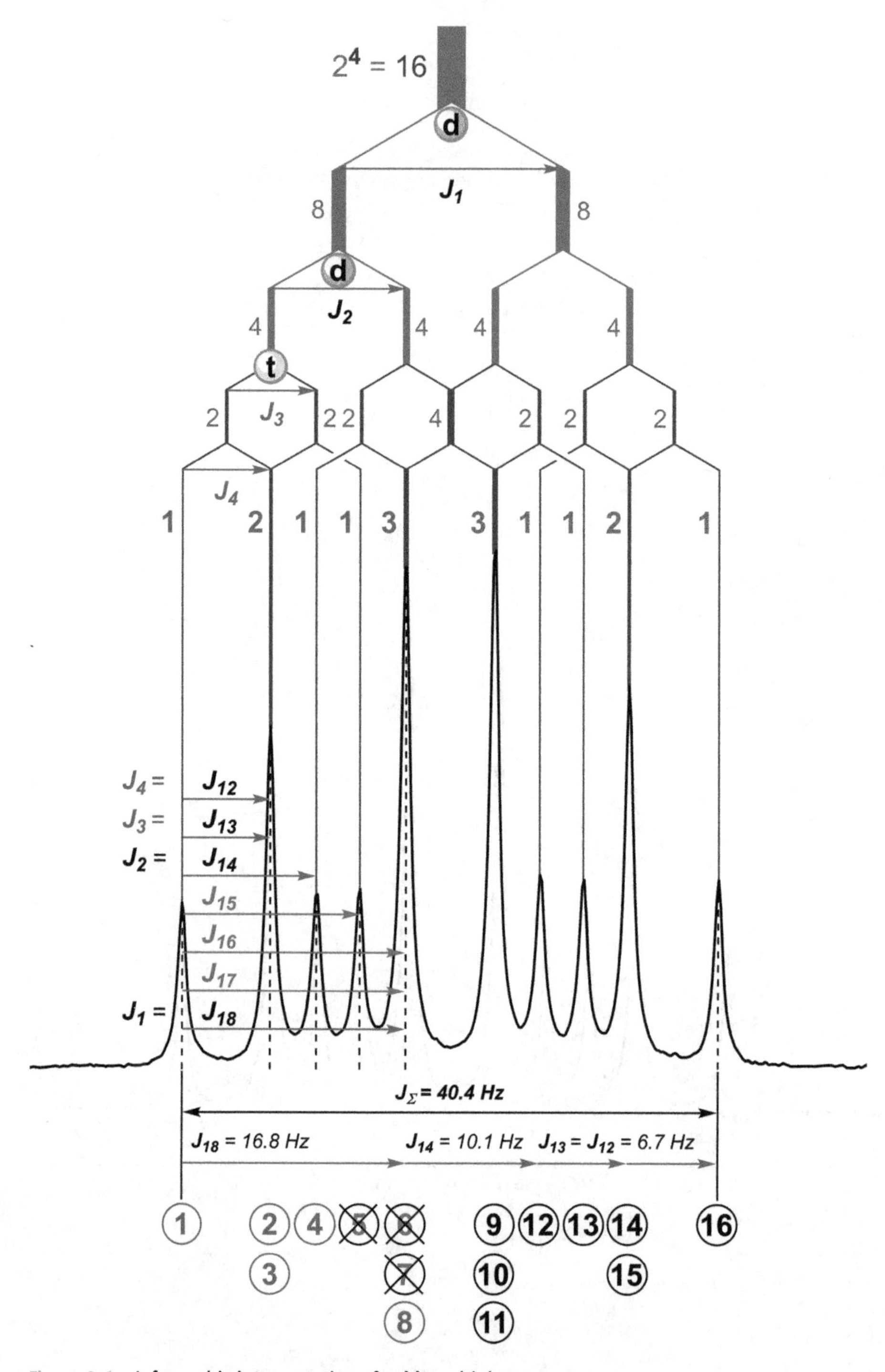

Figure 3.6a: Infographic interpretation of a **ddt** multiplet.

Graphical Representation:

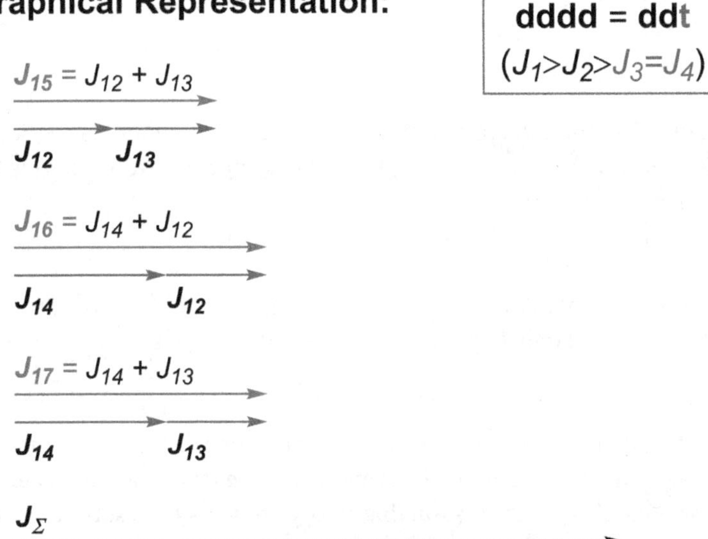

dddd = ddt

$(J_1 > J_2 > J_3 = J_4)$

Numerical Representation:

Measured values:

J_{15} = 13.3 Hz
J_{16} = 16.8 Hz
J_{17} = 16.8 Hz

Calculated values:

$J_{15} = J_{12} + J_{13} = 6.7 + 6.7 = 13.4$ Hz
$J_{16} = J_{14} + J_{12} = 10.1 + 6.7 = 16.8$ Hz
$J_{17} = J_{14} + J_{13} = 10.1 + 6.7 = 16.8$ Hz

Measured sum of all J constants:

$J_{\Sigma} = 40.4$ **Hz**

Calculated sum of all J constants:

$J_{12} + J_{13} + J_{14} + J_{18} = 6.7 + 6.7 + 10.1 + 16.8 = 40.3$ Hz

$J_4 + J_3 + J_2 + J_1 = 6.7 + 6.7 + 10.1 + 16.8 = 40.3$ Hz

Report Representation:

dddd, J = 16.8, 10.1, 6.7, 6.7 Hz
ddt, J = 16.8, 10.1, 6.7 Hz

2^n $(N_J = n)$ $2^4 = 16$ $(N_J = 4)$

Figure 3.6b: Infographic analysis of a **ddt** multiplet.

3.2 Triplet of triplets [tt]

3.2.1 Step 1: Integration

As shown in Figure 3.7, the integration of the first half of the multiplet is determined to be 1:2:1:2:4. Note that the sum of the obtained integral values in this instance should not equal 10 (ten):

$$1+2+1+2+4=10$$

Since the plane of symmetry (σ) divides the middle peak in half, the correct way to calculate this value is to use only ½ (one-half) of the middle signal:

$$1+2+1+2+4 \div 2=1+2+1+2+2=\mathbf{8}$$

Please keep in mind that the relative intensity (= integration) of each peak in the multiplet can be alternatively determined using a quick *qualitative* approach: by comparing the heights of the peaks, assuming the peaks are well-resolved, sharp, and they are not overlapping with other signals (like in this instance).

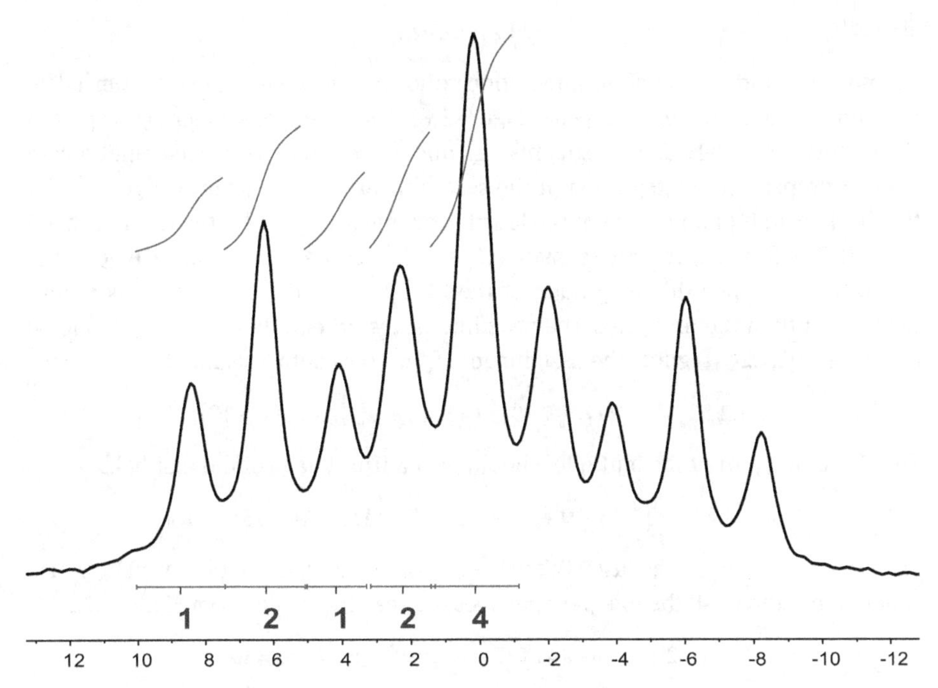

Figure 3.7: Integration of the individual peaks in a **tt** multiplet.

3.2.2 Step 2: Symmetry and 2^n Rule

Because of the *plane of symmetry* (σ), the integration and assignment of the remaining peaks will be a mirror image of the first half (i.e., [4] 2:1:2:1) or:

$$1{:}2{:}1{:}2\,[4]\,2{:}1{:}2{:}1$$

Keeping in mind this relative integration ratio, consecutive *segments* from {1} to {**2^n**} can be placed under each peak (Figure 3.8). For example, segment {1} is placed under the first peak. Both segments {2} and {3} should be placed underneath the second peak (the integration of the second peak is 2). Segment {4} goes under the third peak; {5} and {6} under the fourth (the integral value for the third peak is 2 as well). The final degenerate segments {7} and {8}, as well as {**9**} and {**10**} go under the fifth peak, since the integration of the fifth peak is 4 (four). This peak is four times taller than the first peak. The peak intensities (integration or ≈ height in ideal cases, i.e., 1:2:1:2 [4]) guide the assignment of the consecutive segments:

$$1{:}2{:}1{:}2\,[4] \rightarrow \{1\}, \{2+3\}, \{4\}, \{5+6\}, [7+8+\mathbf{9}+\mathbf{10}]$$

The remaining part of the multiplet should be a mirror image of the first half:

$$[4]\,2{:}1{:}2{:}1 \rightarrow [7+8+\mathbf{9}+\mathbf{10}], \{\mathbf{11}+\mathbf{12}\}, \{\mathbf{13}\}, \{\mathbf{14}+\mathbf{15}\}, \{\mathbf{16}\}$$

The numerical value of the last assigned segment in this case is {**16**} as well. It is also equal to the sum of all the integration values (or peak intensities from Figure 3.7):

$$(1+2+1+2)+(4)+(2+1+2+1)=6+4+6=\mathbf{16}$$

$$\mathbf{16}=\{\mathbf{16}\}$$

According to **Mnemonic Rule I**, introduced in Section 3.1.2, there are **4** (four) true coupling constants associated with this multiplet:

$$\log_2 16=\mathbf{4}$$

$$2^n=16,\ \boldsymbol{n}=\mathbf{4}$$

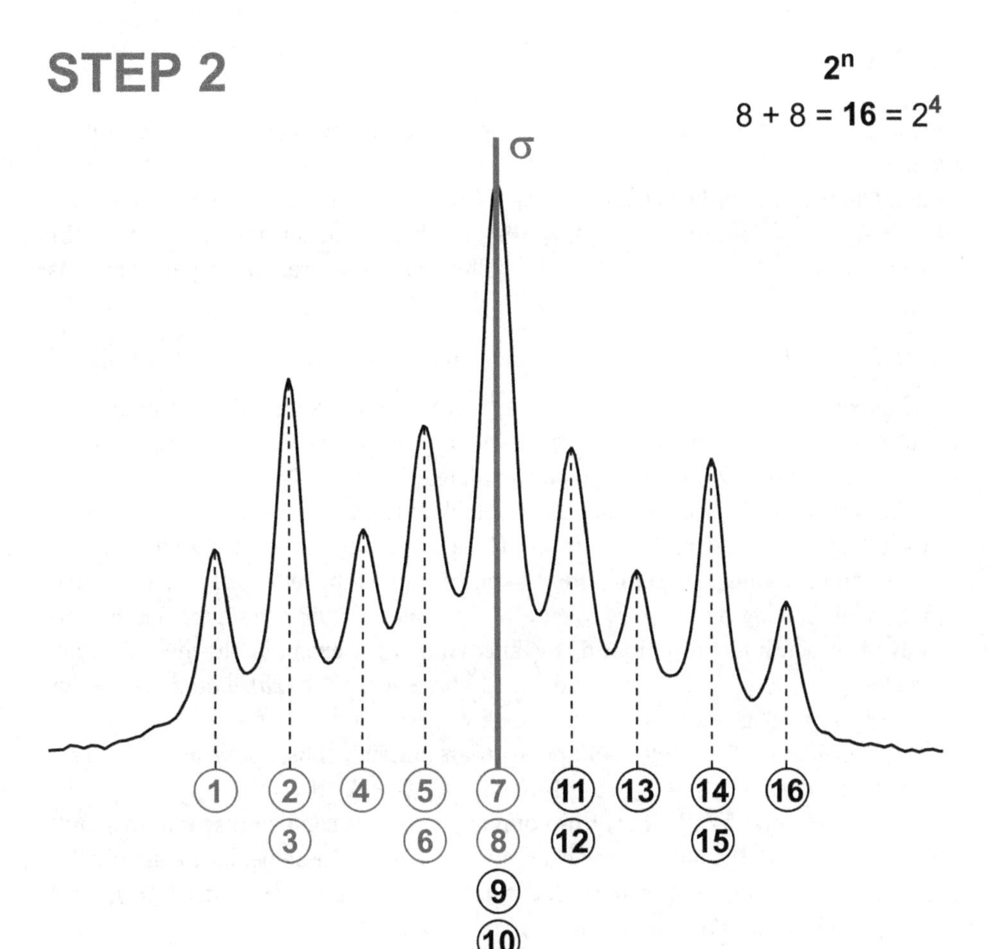

Figure 3.8: Numbering the individual peaks in a **tt** multiplet according to their integration.

3.2.3 Step 3: Coupling Constants

As described previously, the distance between the first peak {1} and the second {2} will always represent the value of the first true coupling constant ($\boldsymbol{J_{1\to2} = J_{12}}$). Since the integral value of the second peak is 2 (two), there should be another constant $\boldsymbol{J_{1\to3} = J_{13}}$, equivalent to J_{12}, that is, $\boldsymbol{J_{12} = J_{13}}$ (both are true constants). Applying the same algorithm, we can identify the following vectors as potential constants (Figure 3.9):

$$J_{1\to4}, J_{1\to5}, J_{1\to6}, J_{1\to7}, J_{1\to8}, J_{1\to9}, J_{1\to10}, J_{1\to11}, J_{1\to12}, J_{1\to13}, J_{1\to14}, J_{1\to15}, \text{ and } J_{1\to16}$$

It is important to remember that not every measured distance will be proportional to a true coupling constant. Linear combinations of the previously determined true coupling constants ($\boldsymbol{J_{12}, J_{13}}$) should be eliminated.

To illustrate this, we can take a closer look at red vector 1→4. The distance between the first peak {1} and the third {4} (1→4) is not unique. An apparent constant $\boldsymbol{J_{1\to4}}$ is merely a linear combination of J_{12} and J_{13}; thus, the length of $J_{1\to4}$ will equal the sum of the length of vectors J_{12} and J_{13}. We can also demonstrate this mathematically by measuring and comparing distances in Hz. For example, the *measured* values for $J_{12} = J_{13} = 2.15$ Hz and $J_{1\to4} = 4.3$ Hz. The *calculated* value for $\boldsymbol{J_{1\to4}} = J_{12} + J_{13} = 2.15 + 2.15 = 4.3$ Hz.

However, the distances 1→5 and 1→6 are unique; thus, these are other true coupling constants: $\boldsymbol{J_{1\to5} = J_{15}} = 6.1$ Hz and $\boldsymbol{J_{1\to6} = J_{16}} = 6.1$ Hz.

Since no more than 4 (four) true coupling constants can be present in this multiplet ($\log_2 16 = 4$), the remaining peaks can be omitted. These peaks are simply linear combinations of the previously identified true coupling constants: $\boldsymbol{J_{12} = J_{13}}$ and $\boldsymbol{J_{15} = J_{16}}$ (*measured value in parentheses*):

$$\boldsymbol{J_{1\to7}} = J_{15} + J_{12} = 6.1 + 2.15 = 8.25\text{ Hz } (\textit{measured as 8.3 Hz})$$

$$\boldsymbol{J_{1\to8}} = J_{15} + J_{13} = 6.1 + 2.15 = 8.25\text{ Hz } (\textit{measured as 8.3 Hz})$$

$$\boldsymbol{J_{1\to9}} = J_{16} + J_{12} = 6.1 + 2.15 = 8.25\text{ Hz } (\textit{measured as 8.3 Hz})$$

$$\boldsymbol{J_{1\to10}} = J_{16} + J_{13} = 6.1 + 2.15 = 8.25\text{ Hz } (\textit{measured as 8.3 Hz})$$

$$\boldsymbol{J_{1\to11}} = J_{15} + J_{12} + J_{13} = 6.1 + 2.15 + 2.15 = 10.4\text{ Hz } (\textit{measured as 10.4 Hz})$$

$$\boldsymbol{J_{1\to12}} = J_{16} + J_{12} + J_{13} = 6.1 + 2.15 + 2.15 = 10.4\text{ Hz } (\textit{measured as 10.4 Hz})$$

$$\boldsymbol{J_{1\to13}} = J_{15} + J_{16} = 6.1 + 6.1 = 12.2\text{ Hz } (\textit{measured as 12.3 Hz})$$

$$\boldsymbol{J_{1\to14}} = J_{15} + J_{16} + J_{12} = 6.1 + 6.1 + 2.15 = 14.35\text{ Hz } (\textit{measured as 14.4 Hz})$$

$$\boldsymbol{J_{1\to15}} = J_{15} + J_{16} + J_{13} = 6.1 + 6.1 + 2.15 = 14.35\text{ Hz } (\textit{measured as 14.4 Hz})$$

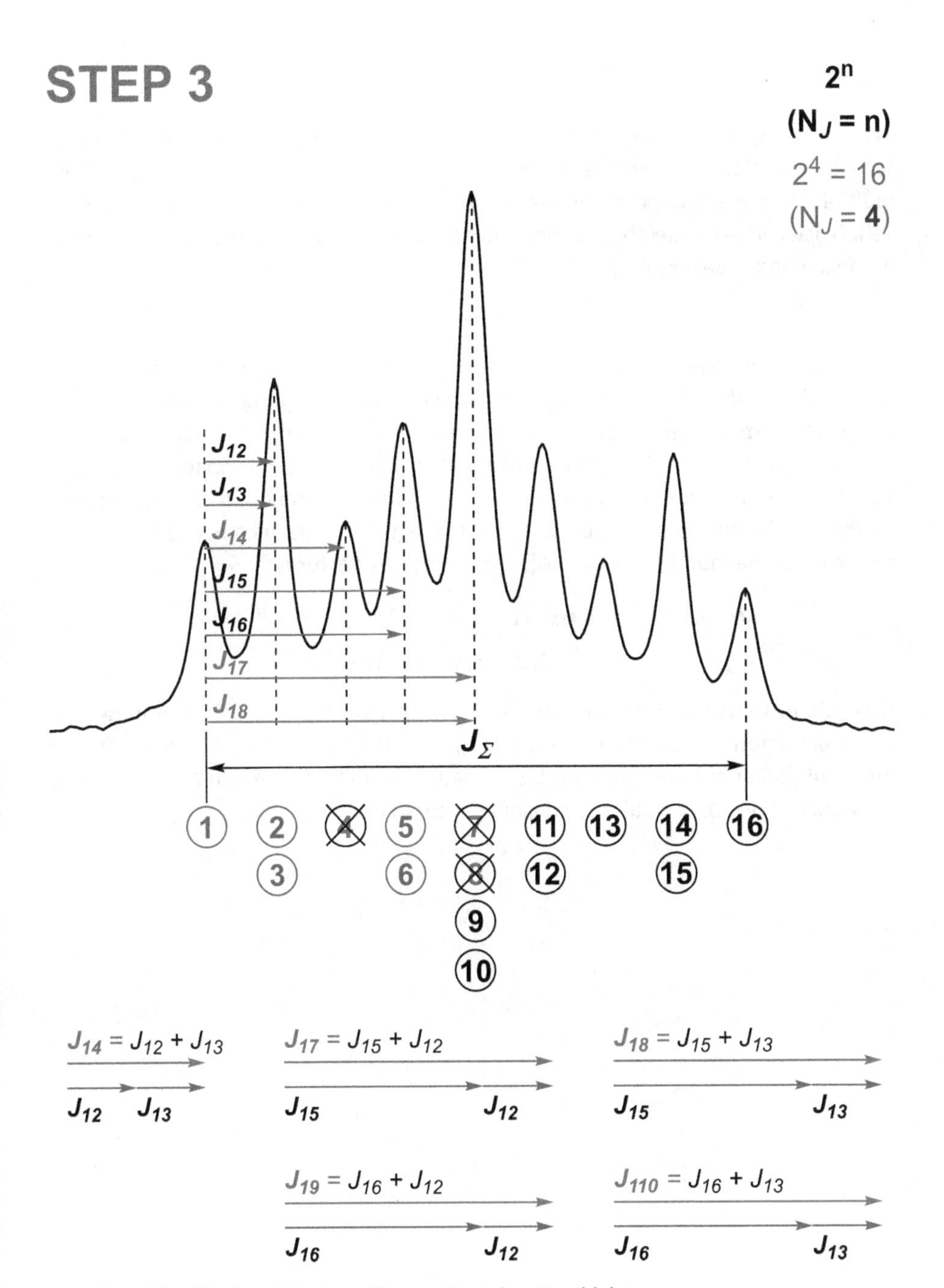

$J_{14} = J_{12} + J_{13}$ (J_{12}, J_{13})

$J_{17} = J_{15} + J_{12}$ (J_{15}, J_{12})

$J_{18} = J_{15} + J_{13}$ (J_{15}, J_{13})

$J_{19} = J_{16} + J_{12}$ (J_{16}, J_{12})

$J_{110} = J_{16} + J_{13}$ (J_{16}, J_{13})

Figure 3.9: Identification of the *J*-coupling constants in a **tt** multiplet.

3.2.4 Step 4: *J*-Tree Diagram

The *J*-tree diagram can be reconstructed according to the previously described protocol (see Section 3.1.4) and References [5, 12]. First, we select the top middle of the multiplet as a starting point (Figure 3.10). The first two arms of the tree should be split twice and separated by the longest vectors (1→5 or 1→6) representing the largest degenerate true coupling constants:

$$J_{15} = J_{16}$$

Since the distance between each branch is identical, this shape (pattern) should be called a *triplet*. The remaining 2 (two) degenerate coupling constants $J_{12} = J_{13}$ complete the tree diagram, that is, each branch of the first *triplet* is split further into three additional *triplets* (or three *doublet of doublets* with equally spaced branches). To follow **Mnemonic Rule III**, we can confirm that the sum of all the blue vectors (or the true *J*-values represented by $\boldsymbol{J_{1\rightarrow16}}$) will equal the distance between the frontier peaks of the multiplet {1} → {**16**}, represented by vector $\boldsymbol{J_\Sigma}$:

$$\boldsymbol{J_{1\rightarrow16}} = \boldsymbol{J_{16}} + \boldsymbol{J_{15}} + \boldsymbol{J_{13}} + \boldsymbol{J_{12}} = 6.1 + 6.1 + 2.15 + 2.15 = 16.5\ \text{Hz}$$

$$\boldsymbol{J_\Sigma} = 16.6\ \text{Hz}\ (\textit{measured})$$

This rule of thumb will help to quickly cross-check if the assigned coupling constants are determined correctly. If the $\boldsymbol{J_{1\rightarrow16}} < \boldsymbol{J_\Sigma}$, then the values of one or several true *J*-coupling constants were **under**estimated. And respectively, if $\boldsymbol{J_{1\rightarrow16}} > \boldsymbol{J_\Sigma}$, then the values of one or several true constants were **over**estimated.

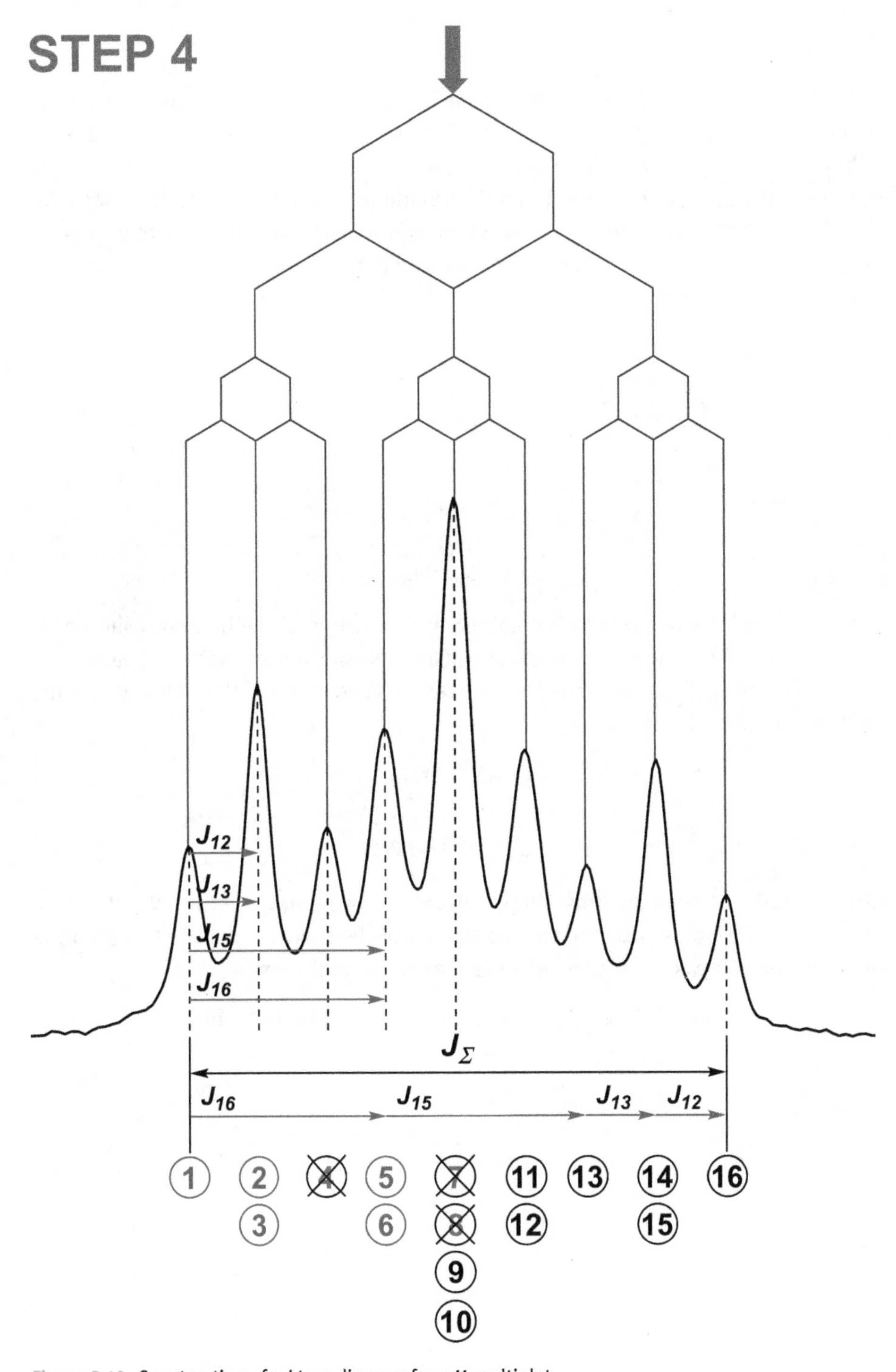

Figure 3.10: Construction of a *J*-tree diagram for a **tt** multiplet.

3.2.5 Step 5: The completed “Puzzle”

As before, we begin with value **16** (sixteen), which we determined using the mnemonic rules described in previous steps: (a) the sum of integral values; (b) $2^n = 2^4 = 16$, where n is the number of J-coupling constants or the number of neighboring H’s the multiplet is coupled to; (c) the numerical value of the last assigned segment {**16**}; and so on. If this value **16** is placed on top of the J-tree diagram and then it is divided following the propagation of the J-tree branches, the resulting pattern emerges (Figure 3.11):

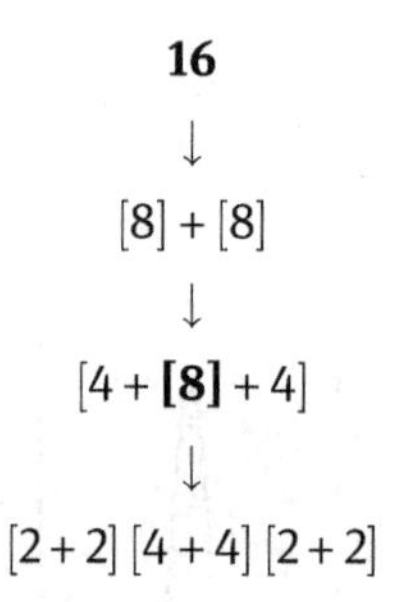

The final products of the values obtained at the bottom of the branches should always equal (⟷) the integration values (peak intensities or ≈ heights) of each peak in the multiplet. Please take into consideration the overlap of the branches for the middle fragment **[8]**:

[2 + 2] [4 + 4] [2 + 2]

↓

1|2|**1**||2|4|**2**||**1**|**2**|**1**

And indeed, by applying this simple algorithm (**Mnemonic Rule IV**), the final products of the values obtained at the bottom of the branches equal (⟷) the integration values (peak intensities) of each peak in the multiplet:

1|2|**1**||2|4|**2**||**1**|**2**|**1** ⟷ 1 + 2 + 1 + 2 + **4** + 2 + 1 + 2 + 1 = **16**

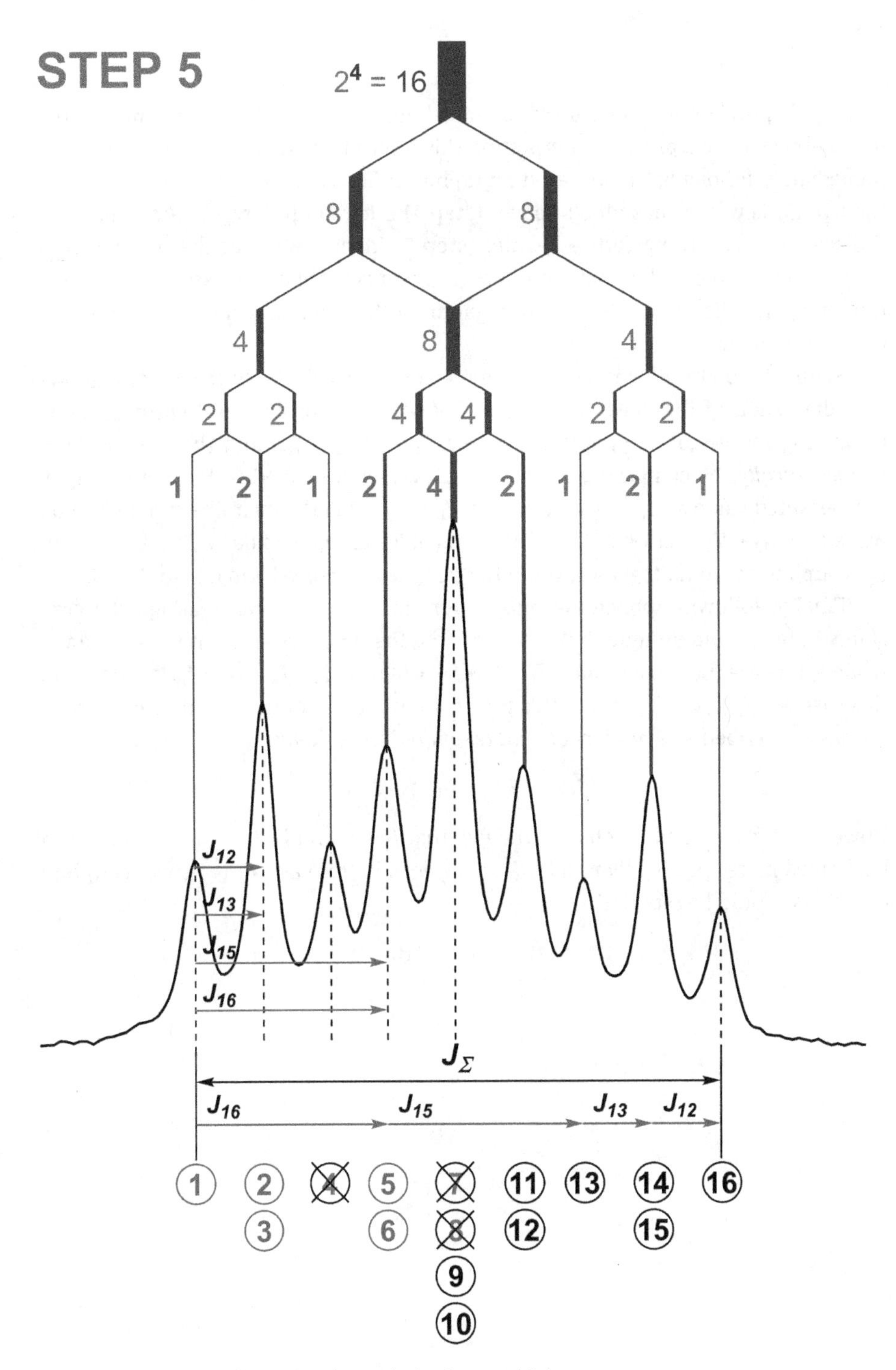

Figure 3.11: Summary illustrating the analysis of a **tt** multiplet.

3.2.6 The Infographic [tt]

Figure 3.12 provides an opportunity to immediately apply the five-step approach detailed above. Please practice interpreting this multiplet and check your work in the immediately following illustrated infographic in Figure 3.12a and b, which covers the five (5) key steps described above: (**step 1**) integration; (**step 2**) assignment of the segments according to the 2^n rule; (**step 3**) identification of the relevant true coupling constants and elimination of their linear combinations; (**step 4**) construction of a *J*-tree diagram; (**step 5**) propagation of the *J*-tree diagram and completion of the assignment.

Figure 3.12b also illustrates **step 3** more comprehensively. In this case, it shows the elimination of the linear combination of coupling constants (a) *graphically*, by comparing the vector lengths, for example, $J_{1\rightarrow4} = J_{12} + J_{13}$, and (b) *numerically = mathematically*, by comparing the constants' values measured in Hz. For example, the measured value for $J_{1\rightarrow4} = 4.3$ Hz. Since $J_{12} = J_{13} = 2.15$ Hz, their linear combination will equal: $J_{12} + J_{13} = 2.15 + 2.15 = 4.3$ Hz. This quick check eliminates $J_{1\rightarrow4}$ as a potential coupling constant, together with other nonrelevant red vectors $J_{1\rightarrow7}$ and $J_{1\rightarrow8}$.

Finally, following prevailing publication guidelines we can reassign the constants notation and assume that $J_1 = J_2$ are the first two largest coupling constants, while $J_3 = J_4$ are the second ones: $\mathbf{J_1} = \mathbf{J_2} = J_{16} = J_{15}$, and $\mathbf{J_3} = \mathbf{J_4} = J_{13} = J_{12}$. Note that in this case $J_1 = J_2 > J_3 = J_4$. Taking this information into account, the multiplet can be formally described as a *doublet of doublet of doublet of doublets*:

$$\textbf{dddd},\ J = 6.1,\ 6.1,\ 2.2,\ 2.2\,\text{Hz}$$

However, to reflect and emphasize the fact that $J_1 = J_2$ and $J_3 = J_4$, the multiplet can be defined more specifically as a *triplet of triplets*. In this case, only 2 (two) coupling constants should be reported:

$$\text{tt},\ J = 6.1,\ 2.2\,\text{Hz}$$

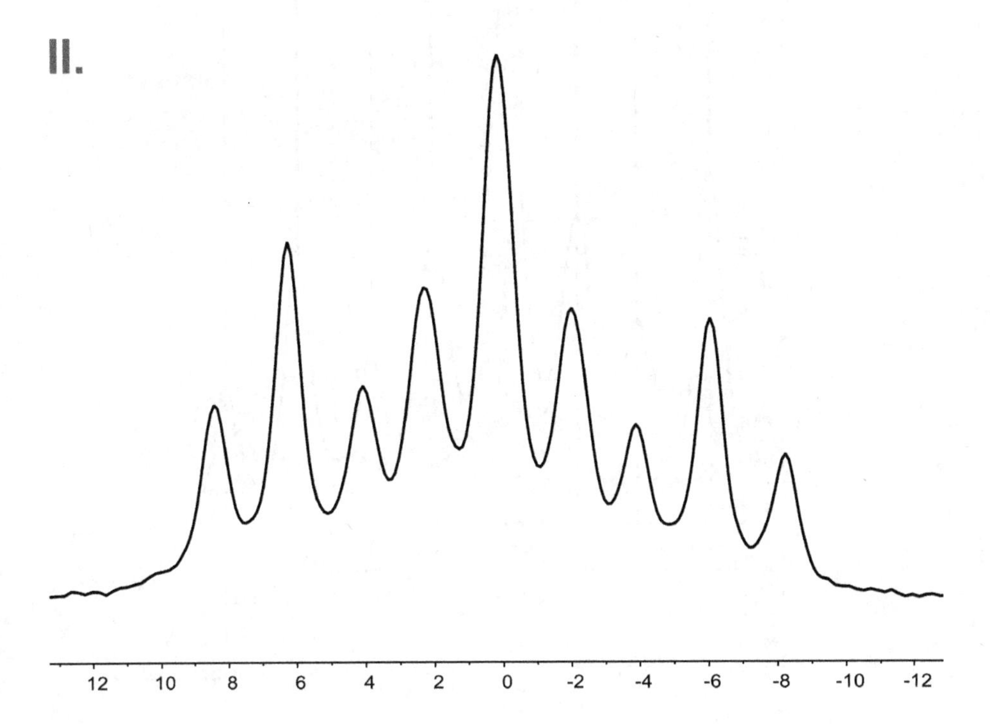

Figure 3.12: Advanced exercise **II** [**tt** multiplet].

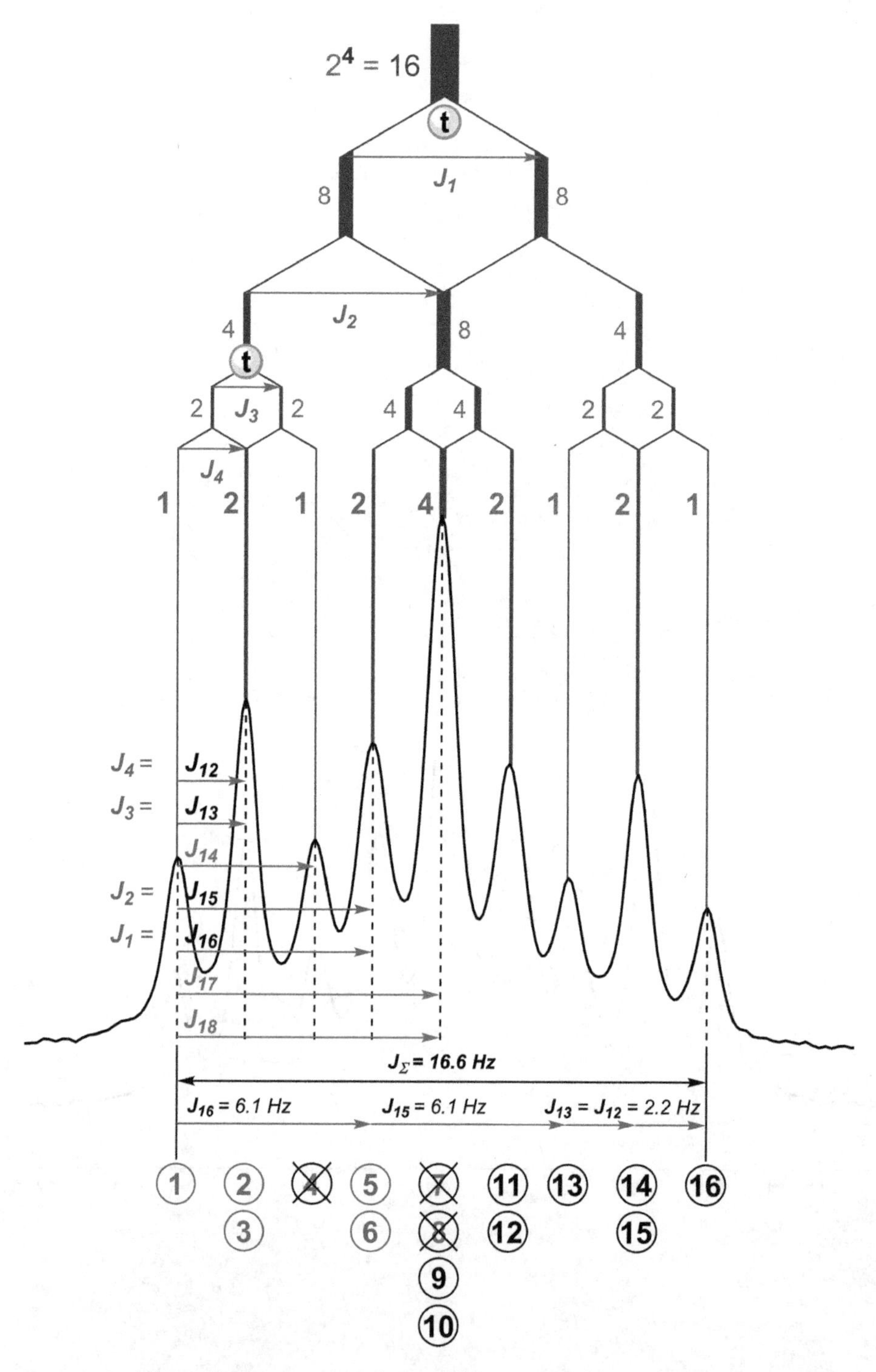

Figure 3.12a: Infographic interpretation of a **tt** multiplet.

Graphical Representation:

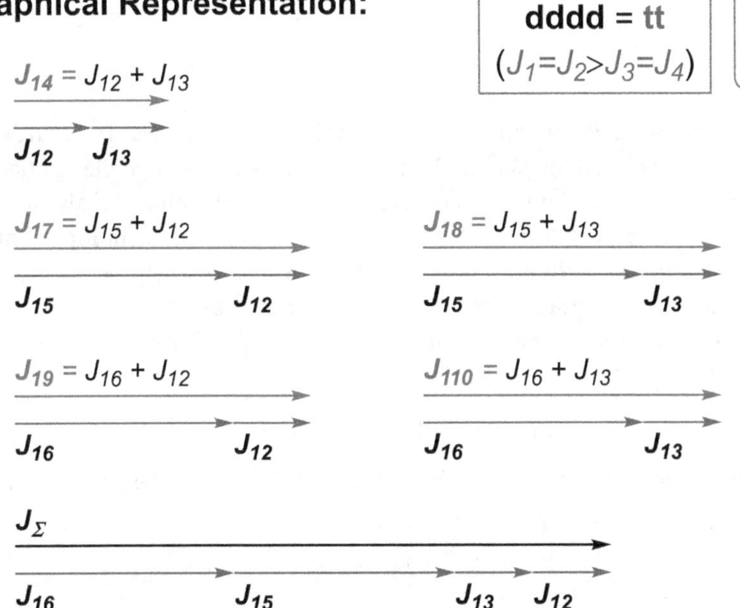

Numerical Representation:

Measured values:	*Calculated values:*
$J_{14} = 4.3$ Hz	$J_{14} = J_{12} + J_{13} = 2.15 + 2.15 = 4.3$ Hz
$J_{17} = 8.3$ Hz	$J_{17} = J_{15} + J_{12} = 6.1 + 2.15 = 8.25$ Hz
$J_{18} = 8.3$ Hz	$J_{18} = J_{15} + J_{13} = 6.1 + 2.15 = 8.25$ Hz

Measured sum of all J constants:

$\mathbf{J_{\Sigma} = 16.6}$ ***Hz***

Calculated sum of all J constants:

$J_{12} + J_{13} + J_{15} + J_{16} = 2.15 + 2.15 + 6.1 + 6.1 = 16.5$ Hz

$J_4 + J_3 + J_2 + J_1 = 2.15 + 2.15 + 6.1 + 6.1 = 16.5$ Hz

Report Representation:

dddd, J = 6.1, **6.1**, 2.2, **2.2** Hz
tt, J = 6.1, 2.2 Hz

2^n ($N_J = n$) $2^4 = 16$ ($N_J = \mathbf{4}$)

Figure 3.12b: Infographic analysis of a **tt** multiplet.

3.3 Doublet of doublet of doublet of doublets [dddd]

3.3.1 Step 1: Integration

This example has a unique feature that can complicate the analysis. The fourth signal (as seen in Figure 3.13) is slightly larger in width but almost similar in height, which is an indication of the presence of two closely overlapping signals. And indeed, as shown in Figure 3.13, the integration of the first half of the multiplet equals 1:1:1:2:1:2. Note that the sixth peak integrates as 2 (two) as well. This is an apparent integration due to the overlap of two closely positioned peaks. Each of those peaks should integrate as 1 (one), however for simplicity we will use the value **2** (two) instead, where [1+1] ≈ [2]:

$$1{:}1{:}1{:}[1+1]{:}1{:}[1+1] \approx 1{:}1{:}1{:}\mathbf{2}{:}1{:}\mathbf{2}$$

The sum of the integral values in this instance is **8** (eight), like in all previous examples:

$$1+1+1+2+1+2=\mathbf{8}$$

Please note that the relative intensity (= integration) of each peak in this multiplet can be properly determined only by using a *quantitative* approach (i.e., integration of individual peaks within the multiplet). A *qualitative* approach based on the comparison of the heights of the peaks is not the best choice due to the overlapping of the signals.

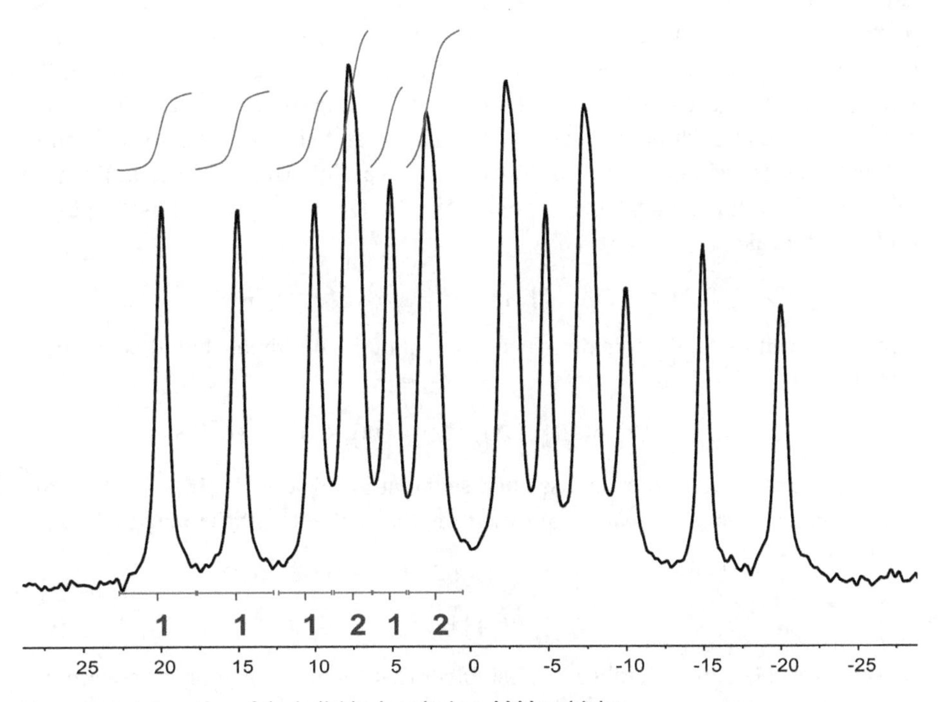

Figure 3.13: Integration of the individual peaks in a **dddd** multiplet.

3.3.2 Step 2: Symmetry and 2^n Rule

Due to the centrosymmetric nature of the multiplet, the remaining integration and assignment of the peaks should be a mirror image of the first half (i.e., 2:1:2:1:1:1) or:

$$1{:}1{:}1{:}2{:}1{:}2\,[\sigma]\,2{:}1{:}2{:}1{:}1{:}1$$

The consecutive *segments* can be placed under each peak according to their integral values (Figure 3.14). Starting with the first segment {1}, which is placed under the first peak, the residual segments {2}, {3}, {4}, . . ., and {8} should be placed under each remaining peak, one digit per one peak of the multiplet. None of the segments in this case are degenerate since none of the signals on their own integrate to 2 (two), 3 (three), or higher. Please note that the apparent integration two (2) (for the fourth and sixth peaks) is a result of the signal overlap. In both cases, the signals are not on top of each other. Instead, they are closely situated next to each other without a clear visible separation. Finally, the peak intensities (1:1:1:2:1:2) can be represented by the sequential segments {1}, {2}, {3}, {4}+{5}, {6}, {7}+{8} placed under each peak:

$$1{:}1{:}1{:}2{:}1{:}2 \rightarrow \{1\}, \{2\}, \{3\}, \{4\}+\{5\}, \{6\}, \{7\}+\{8\}$$

Due to the symmetry, the remaining assignment of the multiplet should be a mirror reflection of the first half:

$$2{:}1{:}2{:}1{:}1{:}1 \rightarrow \{\mathbf{9}\}+\{\mathbf{10}\}, \{\mathbf{11}\}, \{\mathbf{12}\}+\{\mathbf{13}\}, \{\mathbf{14}\}, \{\mathbf{15}\}, \{\mathbf{16}\}$$

The numerical value of the last assigned segment in this case is {**16**}. It equals the sum of all the integration values (representing the peak intensities in Figure 3.13):

$$(1+1+1+2+1+2)+(2+1+2+1+1+1)=\mathbf{8}+\mathbf{8}=\mathbf{16}$$

$$\mathbf{16}=\{\mathbf{16}\}$$

According to **Mnemonic Rule I**, we establish that there are 4 (four) true coupling constants associated with this multiplet:

$$16=2^4,\ \text{or}\ \log_2 16=\mathbf{4}$$

STEP 2

$$2^n$$

$$8 + 8 = \mathbf{16} = 2^4$$

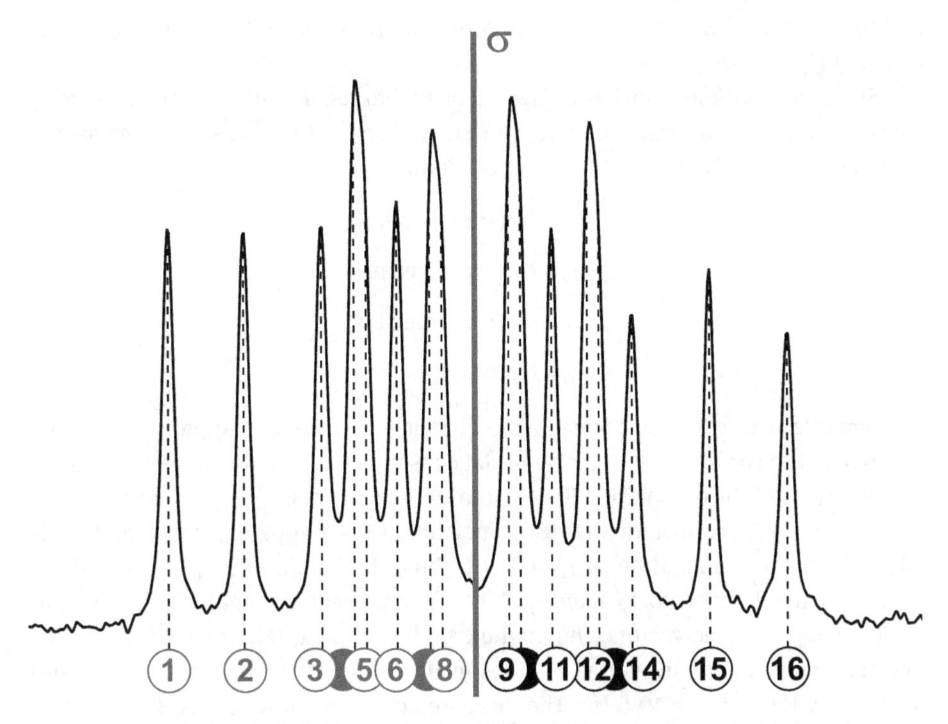

Figure 3.14: Numbering the individual peaks in a **dddd** multiplet according to their integration.

3.3.3 Step 3: Coupling Constants

According to **Mnemonic Rule II**, the distance between the first peak, labeled {1}, and the second {2} will always be proportional to the value of the first true coupling constant ($\boldsymbol{J_{1\to2}}$ or simply $\boldsymbol{J_{12}}$).

Since the multiplet consists of signals of similar height (albeit some are closely overlapping), we can easily determine that the first 4 (four) distances measured from the first peak {1} are true coupling constants:

$$\boldsymbol{J_{1\to2}} = \boldsymbol{J_{12}}\ \text{(true constant)}$$

$$\boldsymbol{J_{1\to3}} = \boldsymbol{J_{13}}\ \text{(true constant)}$$

$$\boldsymbol{J_{1\to4}} = \boldsymbol{J_{14}}\ \text{(true constant)}$$

$$\boldsymbol{J_{1\to5}} = \boldsymbol{J_{15}}\ \text{(true constant)}$$

It is important to carefully measure the distances for the overlapping signals ($\boldsymbol{J_{14}}$ and $\boldsymbol{J_{15}}$) and to confirm that other potential constants (red vectors) are simply linear combinations of the determined first 4 (four) true constants: $\boldsymbol{J_{12}}$, $\boldsymbol{J_{13}}$, $\boldsymbol{J_{14}}$, and $\boldsymbol{J_{15}}$.

And indeed, the distance (1→6) is not unique. An apparent constant $\boldsymbol{J_{1\to6}}$ is merely a linear combination of J_{12} and J_{13}; thus, the length of $J_{1\to6}$ will equal the sum of the length of vectors J_{12} and J_{13}. It can be represented graphically by comparing the lengths of the vectors, the method we just applied (Figure 3.15), or mathematically by using the measured J-values in Hz. For instance, the measured values for J_{12} = 4.9 Hz and J_{13} = 10.0 Hz. The measured distance for $\boldsymbol{J_{1\to6}}$ = 14.85 Hz, which is nearly identical to the calculated value:

$$\boldsymbol{J_{1\to6}} = J_{13} + J_{12} = 10.0 + 4.9 = 14.9\ \text{Hz}\ (\textit{measured as 14.85 Hz})$$

The remaining peaks of the first half of the multiplet can be omitted as well. They are linear combinations of the 4 (four) true coupling constants $\boldsymbol{J_{12}}$, $\boldsymbol{J_{13}}$, $\boldsymbol{J_{14}}$, and $\boldsymbol{J_{15}}$, which we identified previously (*measured value in parenthesis*):

$$\boldsymbol{J_{1\to7}} = J_{14} + J_{12} = 12.2 + 4.9 = 17.1\ \text{Hz}\ (\textit{measured as 17.1 Hz})$$

$$\boldsymbol{J_{1\to8}} = J_{15} + J_{12} = 12.7 + 4.9 = 17.6\ \text{Hz}\ (\textit{measured as 17.6 Hz})$$

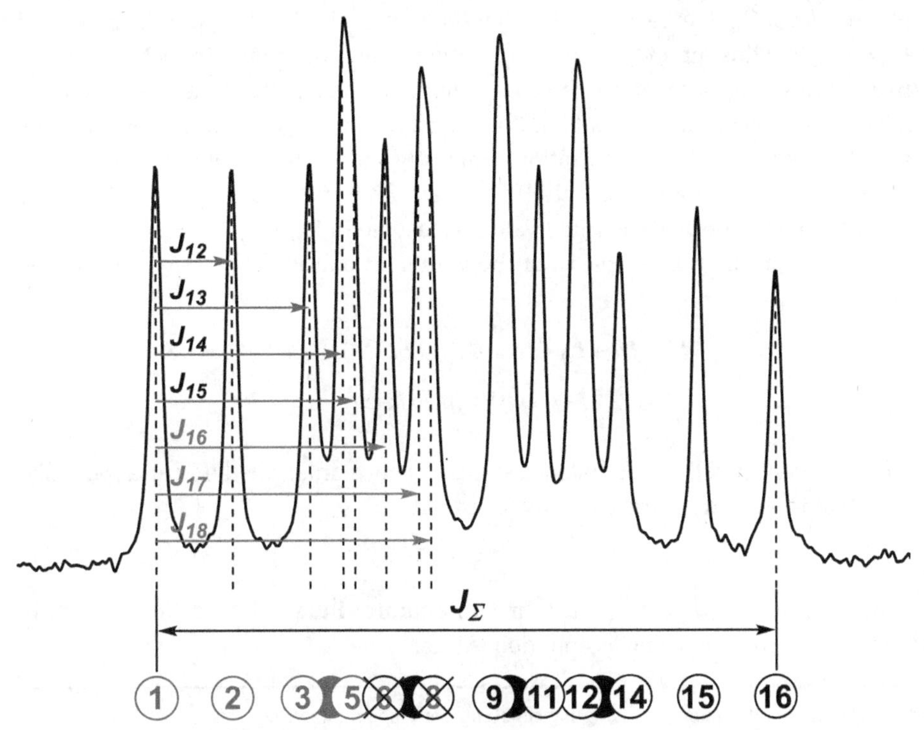

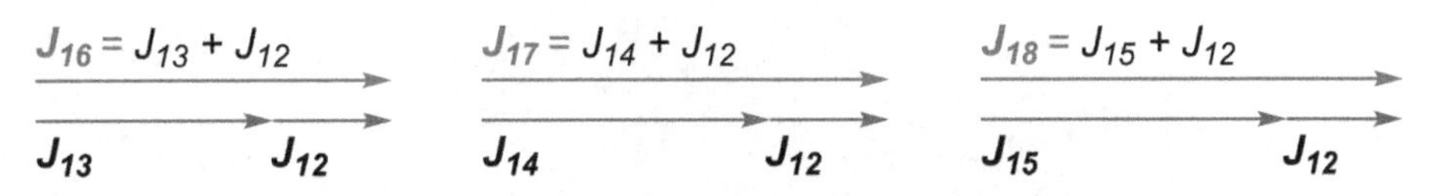

Figure 3.15: Identification of the *J*-coupling constants in a **dddd** multiplet.

3.3.4 Step 4: *J*-Tree Diagram

The *J*-tree diagram can be drawn starting from the top middle of the multiplet (Figure 3.16). The distance between the first two branches should be equal to the largest true coupling constant J_{15} represented by the blue vector 1→5. Since the distance between each branch is unique, the shape can be described as a combination of true *doublets*. Please note that the similarity in value for J_{15} = 12.7 Hz and J_{14} = 12.2 Hz adds an additional complication. The propagating branches of the *J*-tree are very closely separated from one another. This results in the observed overlap of several peaks in the final multiplet. As a result, we have the obvious discrepancy between the relative integration = intensity and the height of the peaks.

According to **Mnemonic Rule III**, we can confirm that the sum of all the blue vectors (which represent the true *J*-values measured in Hz, $\boldsymbol{J_{1\rightarrow16}}$) is equal to the distance between the frontier peaks of the multiplet, that is, {1} → {**16**}, represented buy vector $\boldsymbol{J_{\Sigma}}$:

$$\boldsymbol{J_{1\rightarrow16}} = \boldsymbol{J_{15}} + \boldsymbol{J_{14}} + \boldsymbol{J_{13}} + \boldsymbol{J_{12}} = 12.7 + 12.2 + 10.0 + 4.9 = 39.8\ \text{Hz}$$

$$\boldsymbol{J_{\Sigma}} = 39.9\ \text{Hz}\ (\textit{measured})$$

This also confirms that estimated distances (*J*-constants) for the overlapping signals, that is, $\boldsymbol{J_{15}}$ and $\boldsymbol{J_{14}}$, are correctly measured:

$$\boldsymbol{J_{1\rightarrow16}} = \boldsymbol{J_{\Sigma}}$$

Generally, as it was derived from **Mnemonic Rule III** and mentioned in Section 3.2.4, the following assumption is true:

If $J_{1\rightarrow2^n} < J_{\Sigma}$, then the values of *J*-coupling constants were **under**estimated.
If $J_{1\rightarrow2^n} > J_{\Sigma}$, then the values of *J*-coupling constants were **over**estimated.
If $J_{1\rightarrow2^n} = J_{\Sigma}$, then the values of *J*-coupling constants were **correctly** estimated.

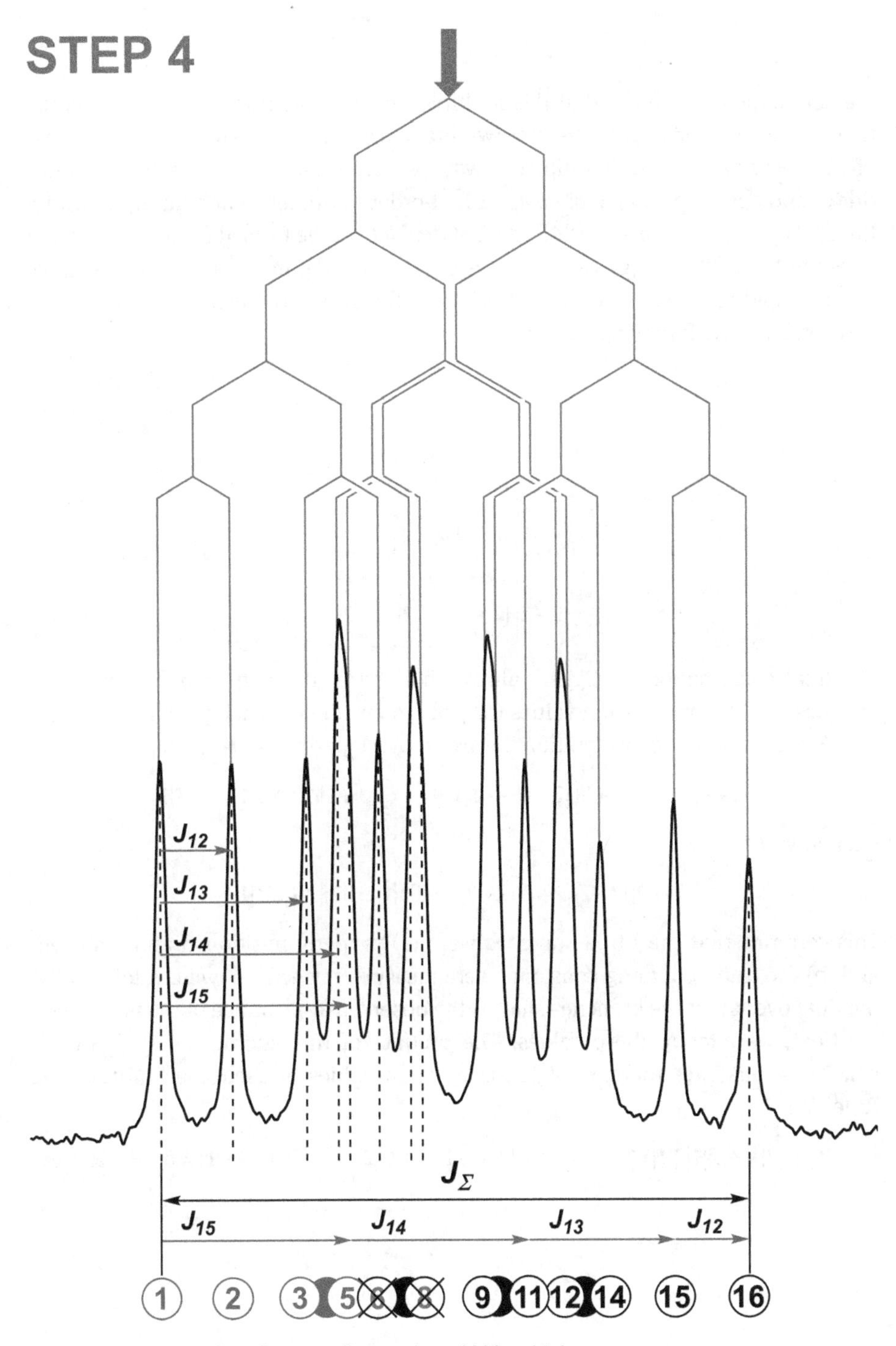

Figure 3.16: Construction of a J-tree diagram for a **dddd** multiplet.

3.3.5 Step 5: The completed "Puzzle"

We determined previously that this multiplet has 4 (four) true coupling constants, thus $2^n = 2^4 = \mathbf{16}$ (sixteen). The first two branches of the *J*-tree diagram, regardless of the complexity of the multiplet, always generate a *doublet*: true for **dd**, **ddd**, **dddd**, and any other *simple* or *complex* first-order multiplets. For a multiplet with 4 (four) coupling constants (**dddd**), the first *doublet* will be formed from the following division: **16** → [8] + [8] (please track this sequence in Figure 3.17). Each consecutive branch should be separated proportionally to the size of the next largest constant; this results in the following pattern:

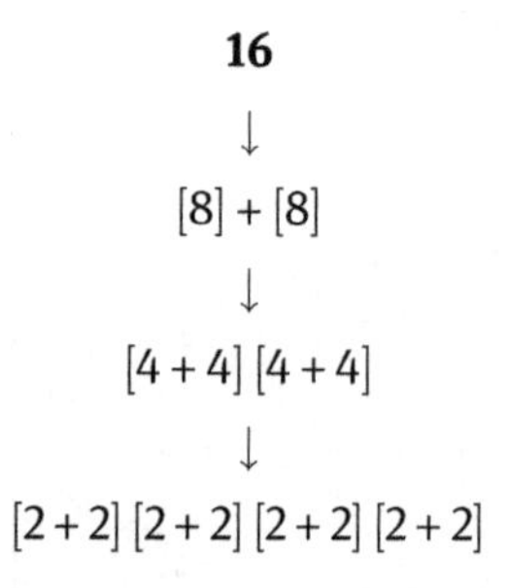

The final combination of digits (values) obtained at the bottom of the branches matches (↔) the integration values (or peak intensities) of each peak in the multiplet. Please take into consideration the signal overlap, that is, [1+1] ≈ [2]:

$$[2+2]\,[2+2]\,[2+2]\,[2+2] \rightarrow 1|1|1|1+1|1|1+1||\mathbf{1}+\mathbf{1}|\mathbf{1}|\mathbf{1}+\mathbf{1}|\mathbf{1}|\mathbf{1}|\mathbf{1}$$

or simply

$$[2+2]\,[2+2]\,[2+2]\,[2+2] \rightarrow \mathbf{1}|\mathbf{1}|\mathbf{1}|\mathbf{2}|\mathbf{1}|\mathbf{2}||\mathbf{2}|\mathbf{1}|\mathbf{2}|\mathbf{1}|\mathbf{1}|\mathbf{1}$$

This confirms that the *J*-tree branches were (a) properly positioned and assigned, and (b) the true *J*-coupling constants were measure correctly as well. Additionally, the final overlap of the branches shaped the observed appearance of the multiplet.

Next, we add up these values. The product of this addition is **16** (sixteen), which also matches the sum of the integration values of each peak (**Mnemonic Rule IV**):

$$1|1|1|2|1|2||\mathbf{2}|\mathbf{1}|\mathbf{2}|\mathbf{1}|\mathbf{1}|\mathbf{1} \leftrightarrow (1+1+1+2+1+2)+(2+1+2+1+1+1)=8+8=\mathbf{16}$$

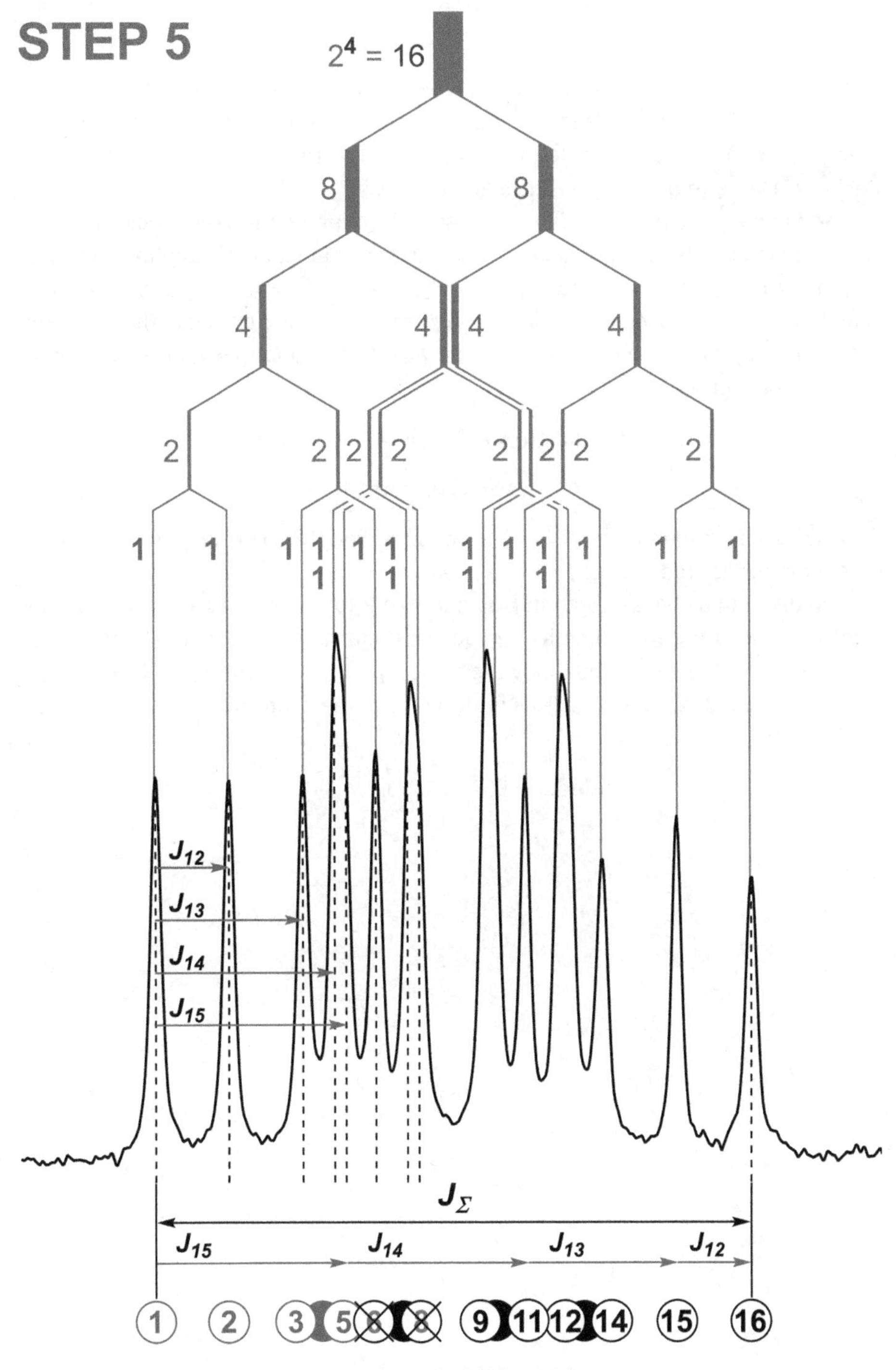

Figure 3.17: Summary illustrating the analysis of a **dddd** multiplet.

3.3.6 The Infographic [dddd]

Figure 3.18 provides an opportunity to immediately apply the five-step approach detailed earlier. Please practice interpreting this multiplet and check your work in the immediately following illustrated infographic in Figure 3.18a and b, which covers the five (5) key assignment steps described above.

Additionally, Figure 3.18b illustrates **step 3** from a mathematical point of view. We can demonstrate the elimination of the linear combination of coupling constants (a) *graphically* (by comparing the vector lengths, e.g., $J_{1\rightarrow 8} = J_{15} + J_{12}$) and (b) *numerically* by comparing the constants' values measured in Hz. For example, the measured value for $J_{1\rightarrow 8} = 17.6$ Hz. Since the values for $J_{15} = 12.7$ Hz and $J_{12} = 4.9$ Hz, their linear combination will equal:

$$J_{1\rightarrow 8}\ (\text{calculated}) = J_{15} + J_{12} = 12.7 + 4.9 = 17.6\ \text{Hz}$$

$$J_{1\rightarrow 8}\ (\text{measured}) = 17.6\ \text{Hz}$$

This quick check eliminates $J_{1\rightarrow 8}$ as a potential coupling constant, together with other nonrelevant red vectors $J_{1\rightarrow 6}$ and $J_{1\rightarrow 7}$.

Finally, let us choose notation $\boldsymbol{J_1}$ and assign it to the largest true coupling constant J_{15}. Comparing the numerical values or vector lengths, the constants can be relabeled and arranged in the following order: $\boldsymbol{J_1} = J_{15}$, $\boldsymbol{J_2} = J_{14}$, $\boldsymbol{J_3} = J_{13}$, $\boldsymbol{J_4} = J_{12}$. Note that in this case $\boldsymbol{J_1} > \boldsymbol{J_2} > \boldsymbol{J_3} > \boldsymbol{J_4}$. This multiplet is a true *doublet of doublet of doublet of doublets*:

dddd, $J = 12.7,\ 12.2,\ 10.0,\ 4.9$ Hz

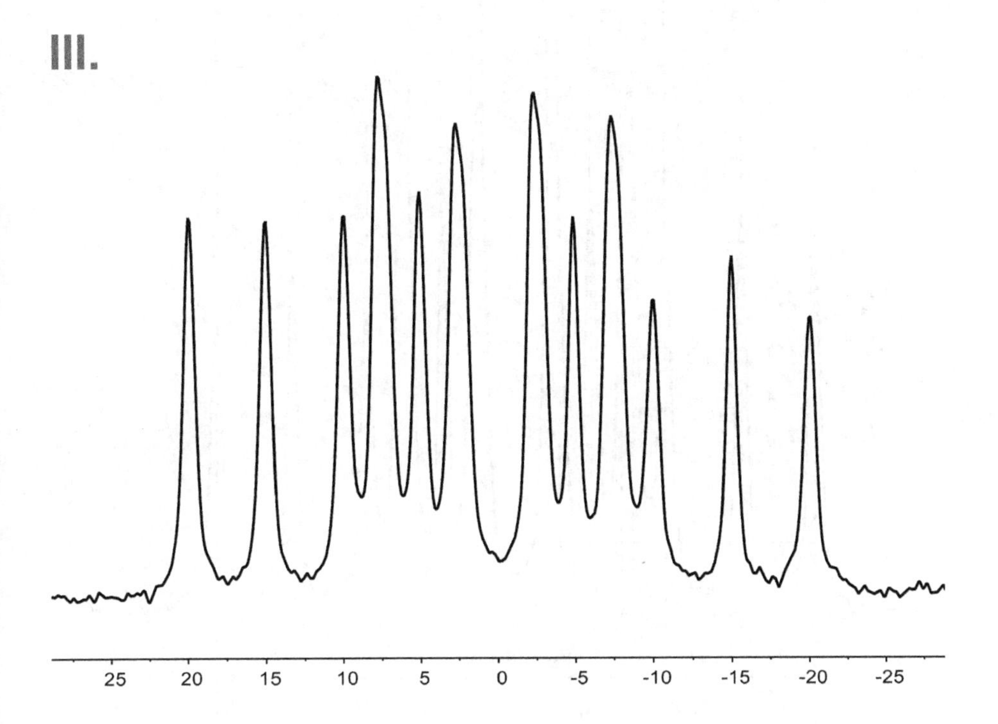

Figure 3.18: Advanced exercise **III** [**dddd** multiplet].

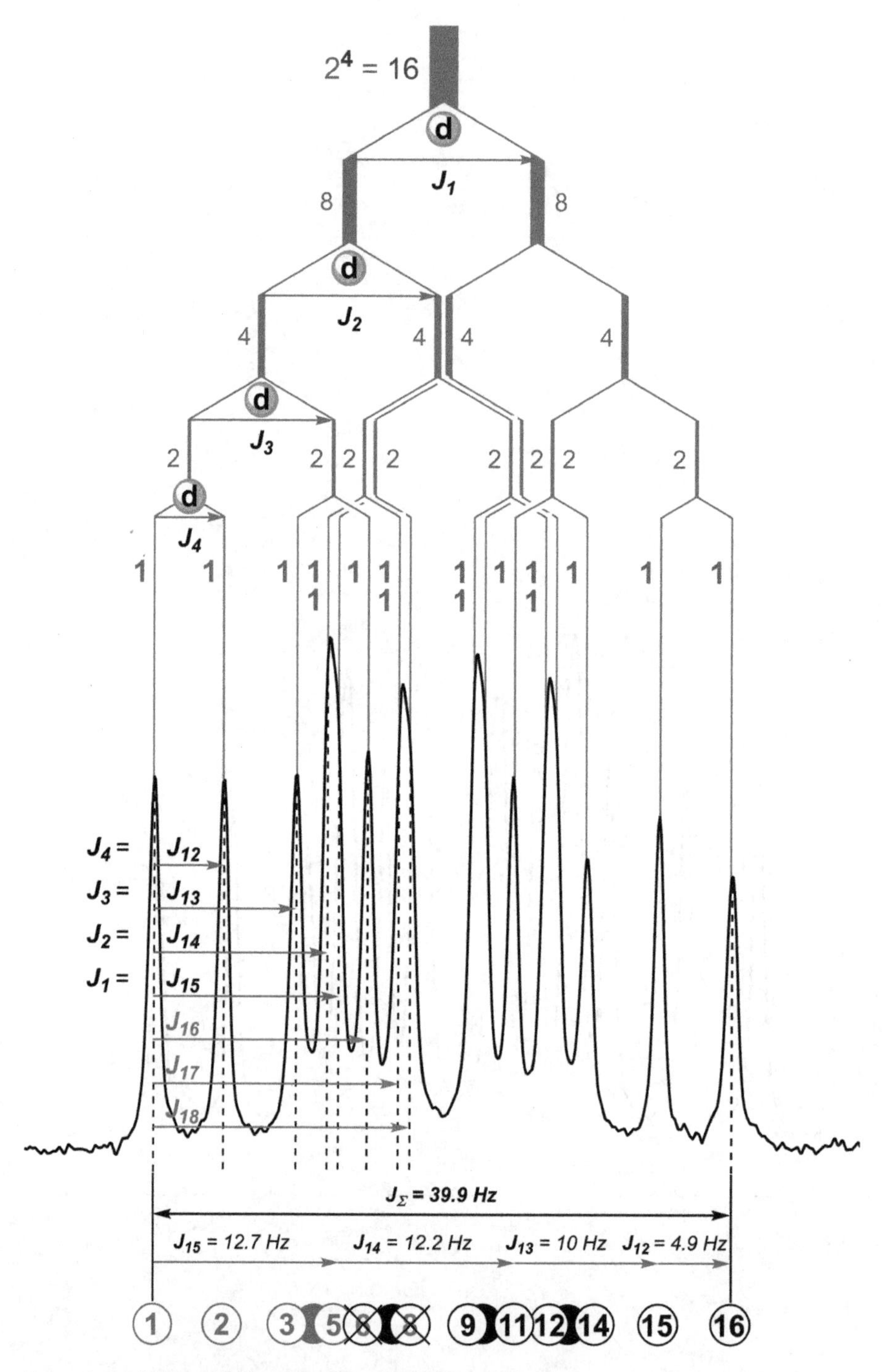

Figure 3.18a: Infographic interpretation of a **dddd** multiplet.

Graphical Representation:

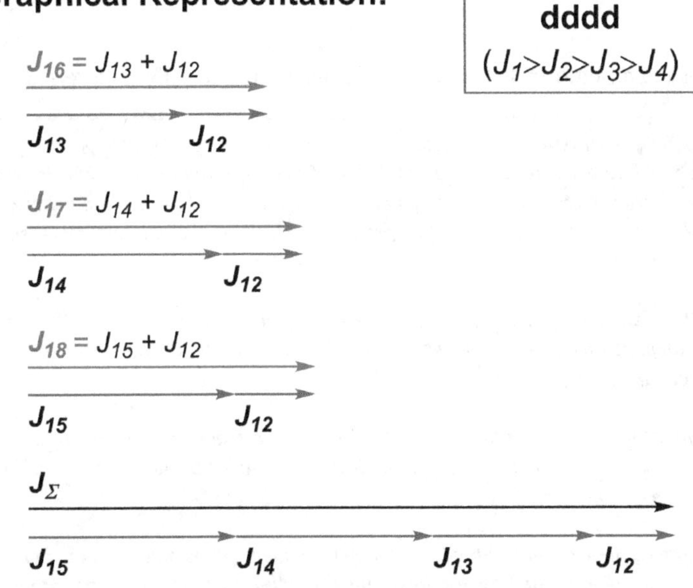

Numerical Representation:

Measured values:	*Calculated values:*
J_{16} = 14.8 Hz	$J_{16} = J_{13} + J_{12} = 10.0 + 4.9 = 14.9$ Hz
J_{17} = 17.1 Hz	$J_{17} = J_{14} + J_{12} = 12.2 + 4.9 = 17.1$ Hz
J_{18} = 17.6 Hz	$J_{18} = J_{15} + J_{12} = 12.7 + 4.9 = 17.6$ Hz

Measured sum of all J constants:

$\boldsymbol{J_{\Sigma} = 39.9}$ **Hz**

Calculated sum of all J constants:

$J_{12} + J_{13} + J_{14} + J_{15} = 4.9 + 10.0 + 12.2 + 12.7 = 39.8$ Hz

$J_4 + J_3 + J_2 + J_1 = 4.9 + 10.0 + 12.2 + 12.7 = 39.8$ Hz

Report Representation:

dddd, J = 12.7, 12.2, 10.0, 4.9 Hz $\qquad 2^n$ (N_J = n) $\qquad 2^4 = 16$ (N_J = **4**)

Figure 3.18b: Infographic analysis of a **dddd** multiplet.

4 Mnemonic rules

Below is the list of the mnemonic rules previously discussed for quick reference.

***Mnemonic Rule I:** (a) For every first-order ^{1}H multiplet, the sum of the integral values of each peak within the multiplet should be equal to the numerical value of the last assigned segment. (b) Note that according to the **2^n rule**: the full sum of the segments always equals **2^n**, **n** is the number of J-coupling constants (N_J) or the number of neighboring protons to which the H of interest, represented by the multiplet, is coupled to.*

***Mnemonic Rule II:** The distance between the very first peak {1} in a first-order ^{1}H multiplet and the second peak {2} (depicted as 1→2) will always equal the value of the smallest true coupling constant in any multiplet ($J_{1\rightarrow 2} = J_{12}$).*

***Mnemonic Rule III:** The distance between the very first peak in a first-order ^{1}H multiplet {1} and the very last one {2^n} will always equal the sum of all true coupling constants in the multiplet ($J_{1\rightarrow 2^n} = J_1 + J_2 + J_3 + \ldots + J_n = J_\Sigma$).*

***Mnemonic Rule IV:** If the numerical value of the last assigned segment {2^n} is placed on top of the J-tree diagram and split according to the propagation of the J-tree branches, then the final products of the values obtained at the bottom of the branches should always equal (⟷) the integral values (peak intensities or ≈ peak heights) of each peak in the multiplet.*

While learning the interpretation of ^{1}H NMR spectra, it can be challenging to see the obvious connection between a **dt** (*doublet of triplets*) and a **q** (*quartet*), for example. The abundance of the examples, the variability in the terminology (**dt, tt, dq, dtd, ttd,** . . ., etc.), and the absence of a systematic approach can further complicate understanding. Table 4.1 helps to visualize the relationship between the number of *J*-coupling constants (***n***), derived from **the 2^n rule**, their magnitude, and it provides a systematic hierarchy illustrating all possible combinations of *Simple First-Order* multiplets: **d, t, q,** and so on.[13] For simplicity and practicality, Table 4.1 summarizes such relationship only for the multiplets with:

1 (one) *J*-coupling constant for a **d** (*doublet*),
2 (two) *J*-coupling constants for a **dd** (*doublet of doublets*),
3 (three) *J*-coupling constants for a **ddd** (*doublet of doublet of doublets*),
4 (four) *J*-coupling constants for a **dddd** (*doublet of doublet of doublet of doublets*),
5 (five) *J*-coupling constants for a **ddddd** (*doublet of doublet of doublet of doublet of doublets*).

13 Complex First-Order multiplets can be viewed as an overlapping combination of Simple First-Order multiplets: **dtd** multiplet is a combination of a *doublet* (**d**), a *triplet* (**t**), and another *doublet* (**d**).

https://doi.org/10.1515/9783110608403-005

Table 4.1: Possible combination of various first-order ^{1}H multiplets with 0, 1, 2, 3, 4, and 5 *J*-coupling constants[a].

$2^0 = 1$	$2^1 = 2$	$2^2 = 4$	$2^3 = 8$	$2^4 = 16$	$2^5 = 32$
0	1	2	3	4	5
s[b]	d[b]	**dd**	**ddd**	**dddd**	**ddddd**
	J_1	$J_1 > J_2$	$J_1 > J_2 > J_3$	$J_1 > J_2 > J_3 > J_4$	$J_1 > J_2 > J_3 > J_4 > J_5$
		t[b]	**td**	**tdd**	**tddd**
		$J_1 = J_2$	$J_1 = J_2 > J_3$	$J_1 = J_2 > J_3 > J_4$	$J_1 = J_2 > J_3 > J_4 > J_5$
			dt	**dtd**	**dtdd**
			$J_1 > J_2 = J_3$	$J_1 > J_2 = J_3 > J_4$	$J_1 > J_2 = J_3 > J_4 > J_5$
			q[b]	**ddt**	**ddtd**
			$J_1 = J_2 = J_3$	$J_1 > J_2 > J_3 = J_4$	$J_1 > J_2 > J_3 = J_4 > J_5$
				tt	**dddt**
				$J_1 = J_2 > J_3 = J_4$	$J_1 > J_2 > J_3 > J_4 = J_5$
				qd	**ttd**
				$J_1 = J_2 = J_3 > J_4$	$J_1 = J_2 > J_3 = J_4 > J_5$
				dq	**tdt**
				$J_1 > J_2 = J_3 = J_4$	$J_1 = J_2 > J_3 > J_4 = J_5$
				quintet[b]	**dtt**
				$J_1 = J_2 = J_3 = J_4$	$J_1 > J_2 = J_3 > J_4 = J_5$
					qdd
					$J_1 = J_2 = J_3 > J_4 > J_5$
					dqd
					$J_1 > J_2 = J_3 = J_4 > J_5$
					ddq
					$J_1 > J_2 > J_3 = J_4 = J_5$
					tq
					$J_1 = J_2 > J_3 = J_4 = J_5$
					qt
					$J_1 = J_2 = J_3 > J_4 = J_5$
					quintet of d
					$J_1 = J_2 = J_3 = J_4 > J_5$
					d of quintets
					$J_1 > J_2 = J_3 = J_4 = J_5$
					sextet[b]
					$J_1 = J_2 = J_3 = J_4 = J_5$

[a]The assumption is that the value of the constants decrease in the order: $J_1 > J_2 > J_3 > J_4 > \ldots > J_n$.
[b]*Simple First-Order Multiplets*: **singlet, doublet, triplet, quartet, quintet, sextet.** Remember that *Fundamental Simple First-Order Multiplets* have a unique single letter notation: singlet (**s**), doublet (**d**), triplet (**t**), quartet (**q**).

Let us consider a multiplet with 3 (three) nonequivalent *J*-coupling constants ($J_1 > J_2 > J_3$, $2^3 = 8$). Such multiplet can only be described as a **ddd** (*doublet of doublet of doublets*). In more specific cases, for example, when $J_1 = J_2 > J_3$, it should be called a **td** (*triplet of doublets*) or a **dt** (*doublet of triplets*) if $J_1 > J_2 = J_3$. Finally, if every single constant is equivalent to one another ($J_1 = J_2 = J_3$), the multiplet should be called a *quartet* (**q**): it is one of the *Fundamental Simple First-Order* multiplets. This systematic visual approach is powerful and can help us recognize that a *quartet* is not a unique type of the multiplets, but merely an example of a *doublet of doublet of doublets*, in which all 3 (three) *J*-coupling constants are identical to one another.

In another example, one can easily recognize that a **ttd** (*triplet of triplet of doublets*) is a more specific case of a **ddddd** (*doublet of doublet of doublet of doublet of doublets*). In this example, the magnitude of the 5 (five) constants are related to one another in the following manner: $J_1 = J_2 > J_3 = J_4 > J_5$. This nomenclature relies on the assumption that J_1 is always the largest constant for any first-order ^{1}H multiplet: $J_1 > J_2 > J_3 > J_4 > \ldots > J_n$.

The previously described five-step procedure is very useful for learning purposes, but is not always time efficient. Once the general principle is well understood based on those examples, we can rely upon the rules of thumb to apply it more efficiently to routine interpretation scenarios. Let us refer to **step 2** mentioned in Chapter 3. If the irrelevant coupling constants are identified and eliminated and only true constants are labeled (represented by the blue circles), the multiplet then can be interpreted graphically, based on the arranged shape of the blue circles. Below is the summary of this suggested mnemonic rule (Figure 4.1).

Mnemonic Rule V: Assume (a) one blue circle represents a **doublet**; (b) two circles (stacked on top of each other) are equivalent to a **triplet**; (c) three circles will represent a **quartet**; and so on (d) the first assigned segment {1} should be always omitted. By starting from the largest constant (largest distance between the first peak and the peak of interest under which the last blue circle is placed) and moving right to left (←), we can visualize a unique geometrical pattern. This arrangement of the circles helps to identify the multiplet type.

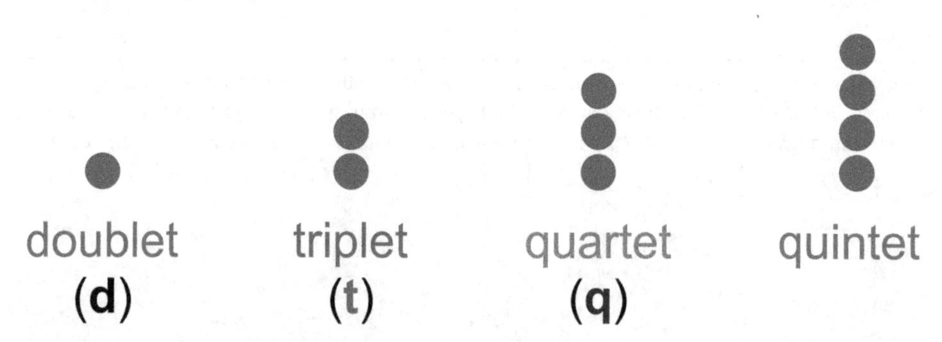

Figure 4.1: Mnemonic Rule V.

For an illustration of this rule, please refer to Figure 4.2. The first example (A) is a general **dddd** (*doublet of doublet of doublet of doublets*) multiplet. The inverse T arrangement of the circles in example (B) corresponds with a **dtd** (*doublet of triplet of doublets*) interpretation. Square example (C) represents a **tt** (*triplet of triplets*). And finally, the L-shape in example (D) can be described as a **dq** (*doublet of quartets*).

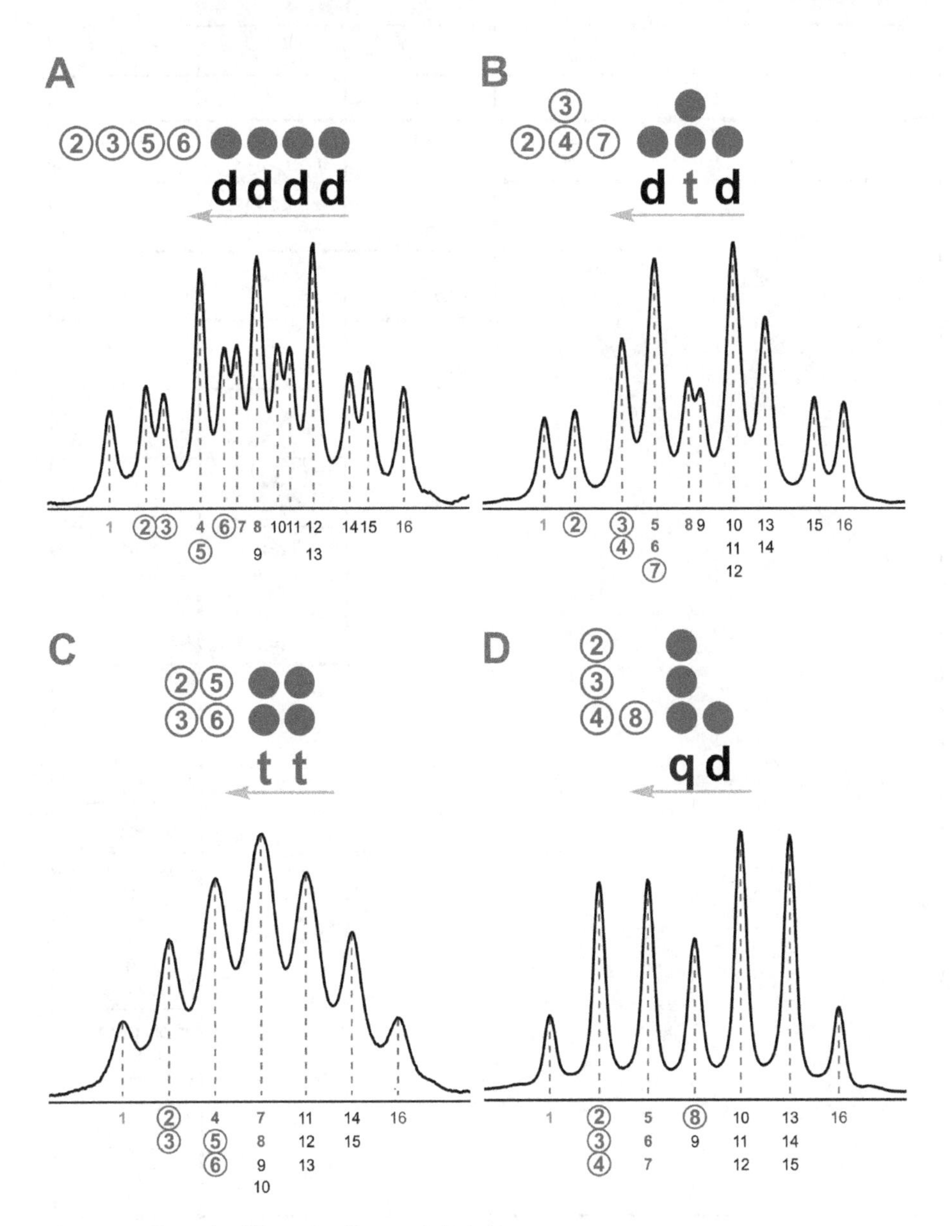

Figure 4.2: Examples illustrating Mnemonic Rule V.

All examples demonstrating **Mnemonic Rule V** are also summarized in Figure 4.4. This is a graphical representation of Figure 4.3 and Table 4.1 and it is meant to be used as a reference for the beginner to quickly assess whether the multiplet of interest was properly assigned and interpreted.

$2^0=1$	$2^1=2$	$2^2=4$	$2^3=8$	$2^4=16$	$2^5=32$
0	1	2	3	4	5
s	d J_1	dd $J_1>J_2$	ddd $J_1>J_2>J_3$	dddd $J_1>J_2>J_3>J_4$	ddddd $J_1>J_2>J_3>J_4>J_5$
		t $J_1=J_2$	td $J_1=J_2>J_3$	tdd $J_1=J_2>J_3>J_4$	tddd $J_1=J_2>J_3>J_4>J_5$
			dt $J_1>J_2=J_3$	dtd $J_1>J_2=J_3>J_4$	dtdd $J_1>J_2=J_3>J_4>J_5$
			q $J_1=J_2=J_3$	ddt $J_1>J_2>J_3=J_4$	ddtd $J_1>J_2>J_3=J_4>J_5$
				tt $J_1=J_2>J_3=J_4$	dddt $J_1>J_2>J_3>J_4=J_5$
				qd $J_1=J_2=J_3>J_4$	ttd $J_1=J_2>J_3=J_4>J_5$
				dq $J_1>J_2=J_3=J_4$	tdt $J_1=J_2>J_3>J_4=J_5$
				quintet $J_1=J_2=J_3=J_4$	dtt $J_1>J_2=J_3>J_4=J_5$
					qdd $J_1=J_2=J_3>J_4>J_5$
					dqd $J_1>J_2=J_3=J_4>J_5$
					ddq $J_1>J_2>J_3=J_4=J_5$
					tq $J_1=J_2>J_3=J_4=J_5$
					qt $J_1=J_2=J_3>J_4=J_5$
					d of quintets $J_1>J_2=J_3=J_4=J_5$
					quintet of d $J_1=J_2=J_3=J_4>J_5$
					sextet $J_1=J_2=J_3=J_4=J_5$

Figure 4.3: Multiplet classification with n (2^n) J-coupling constants.

$2^0=1$	$2^1=2$	$2^2=4$	$2^3=8$	$2^4=16$	$2^5=32$
0	1	2	3	4	5
s	d	dd	ddd	dddd	ddddd
	doublet	t	td	tdd	tddd
		triplet	dt	dtd	dtdd
			q	ddt	ddtd
			quartet	tt	dddt
				qd	ttd
				dq	tdt
					dtt
				quintet	qdd
					dqd
					ddq
					tq
					qt
					d of quintets
					quintet of d
				sextet	

Figure 4.4: Graphic summary of Mnemonic Rule V.

5 Exercises

5.1 Introductory Level: doublet of doublets [dd]

$2^0=1$	$2^1=2$	$2^2=4$	$2^3=8$	$2^4=16$	$2^5=32$
0	1	2	3	4	5
s	d	dd	ddd	dddd	ddddd
	doublet	t	td	tdd	tddd
		triplet	dt	dtd	dtdd
			q	ddt	ddtd
			quartet	tt	dddt
				qd	ttd
				dq	tdt
					dtt
				quintet	qdd
					dqd
					ddq
					tq
					qt
					d of quintets
					quintet of d
				sextet	

Figure 5.1: Multiplets with two J-coupling constants ($n = 2$: J_1, J_2).

https://doi.org/10.1515/9783110608403-006

5.1.1 Introductory Exercises

1.

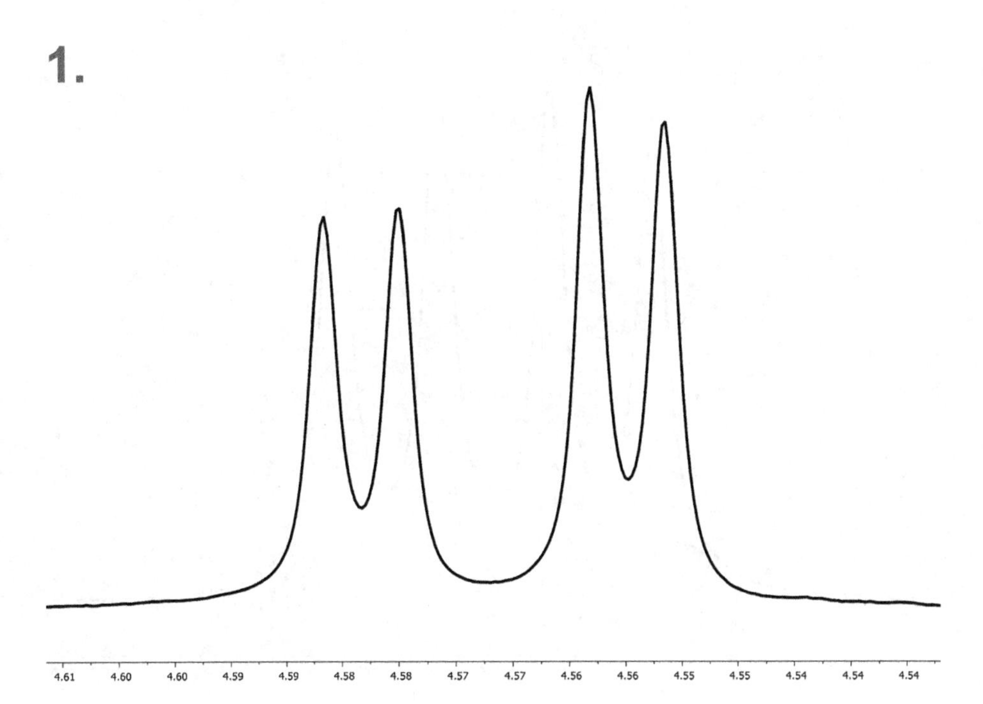

Figure 5.2: Introductory exercise 1.

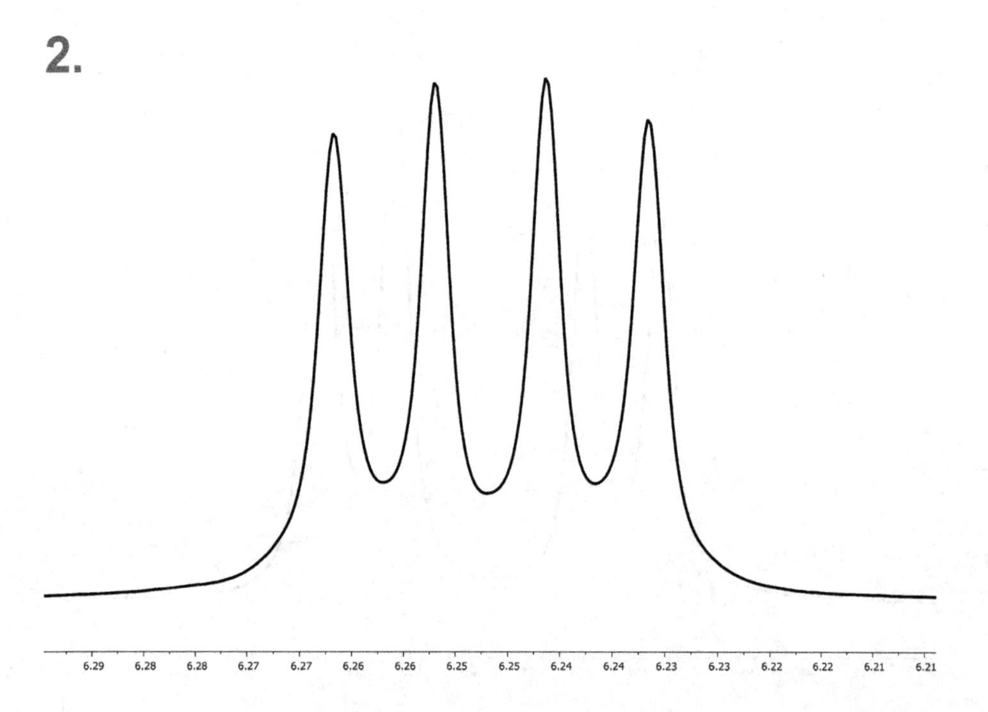

Figure 5.3: Introductory exercise 2.

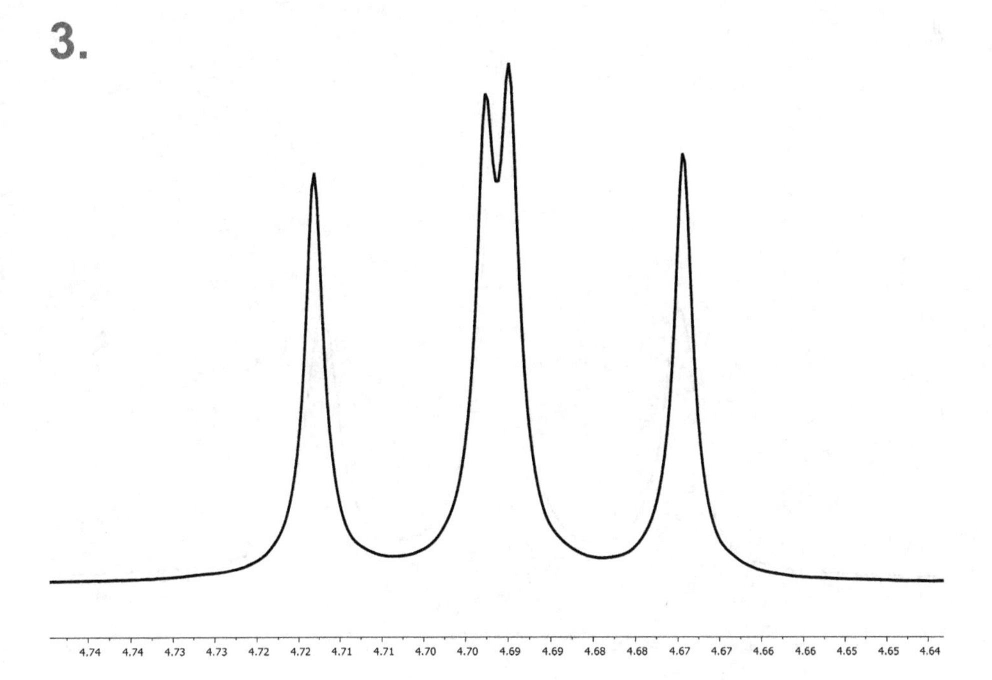

Figure 5.4: Introductory exercise 3.

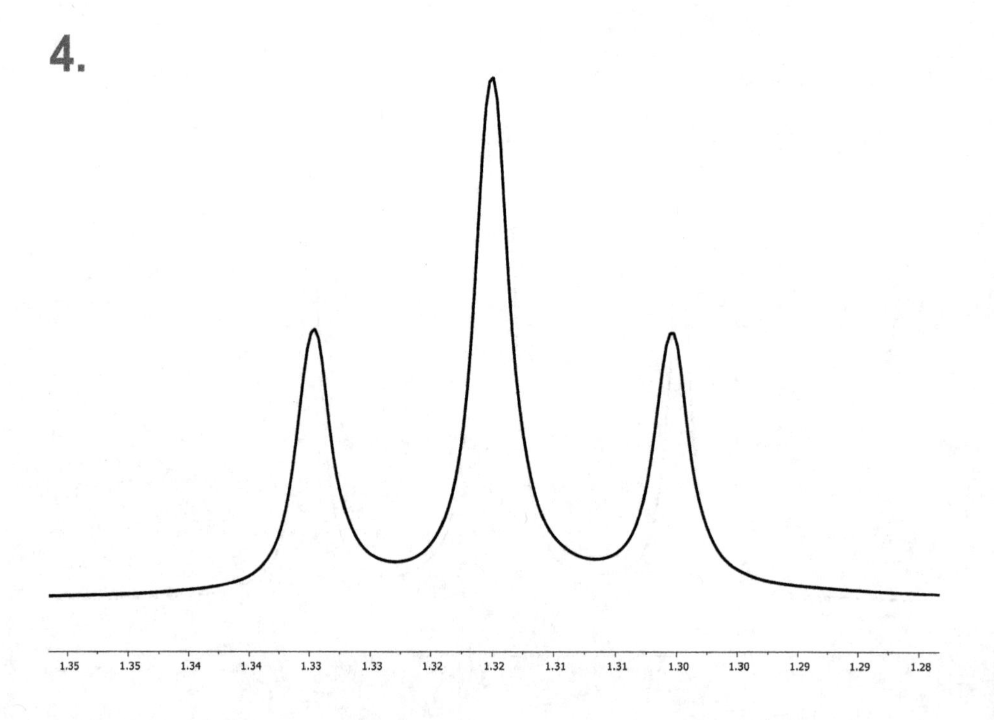

Figure 5.5: Introductory exercise 4.

5.1.2 Introductory Answers

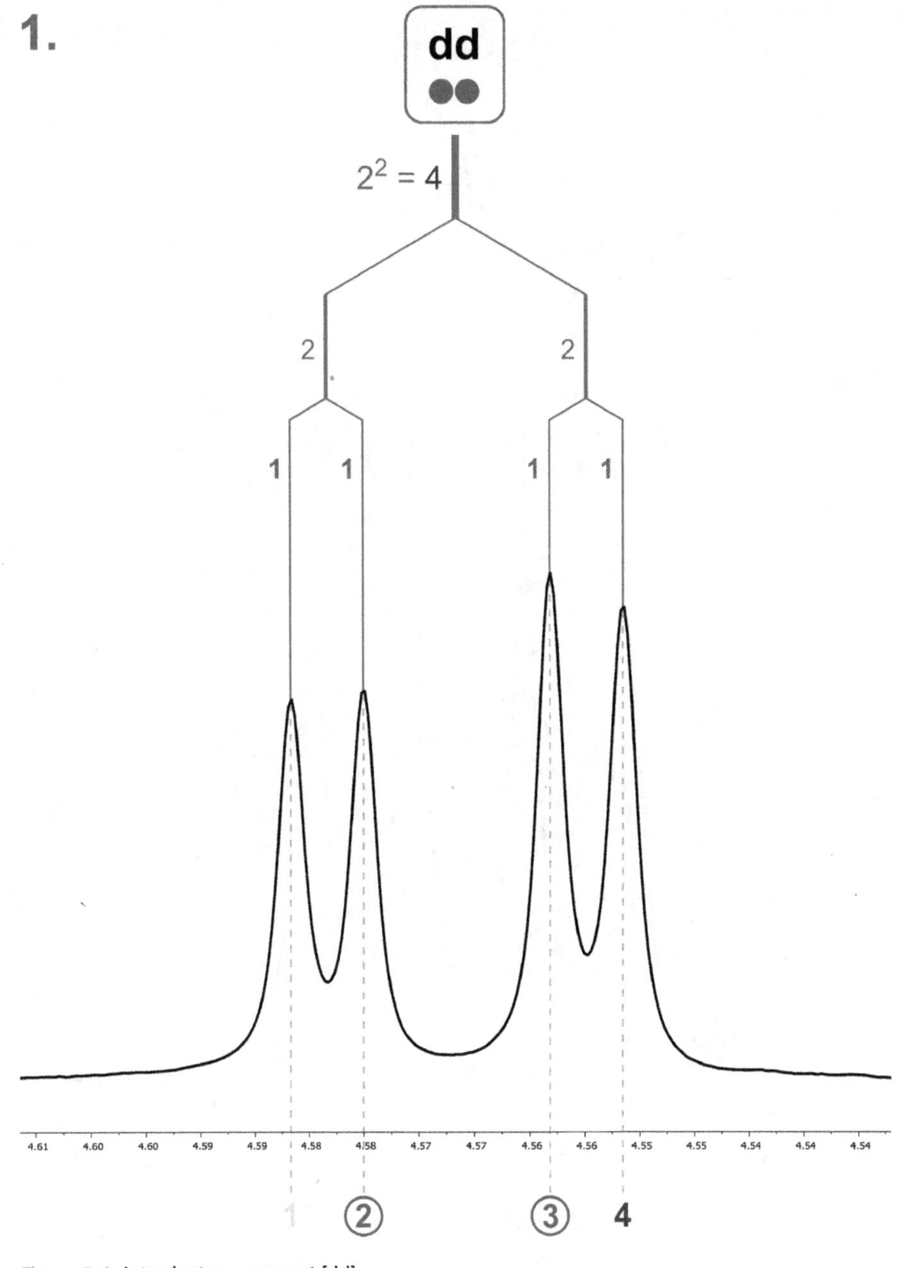

Figure 5.6: Introductory answer 1 [dd].

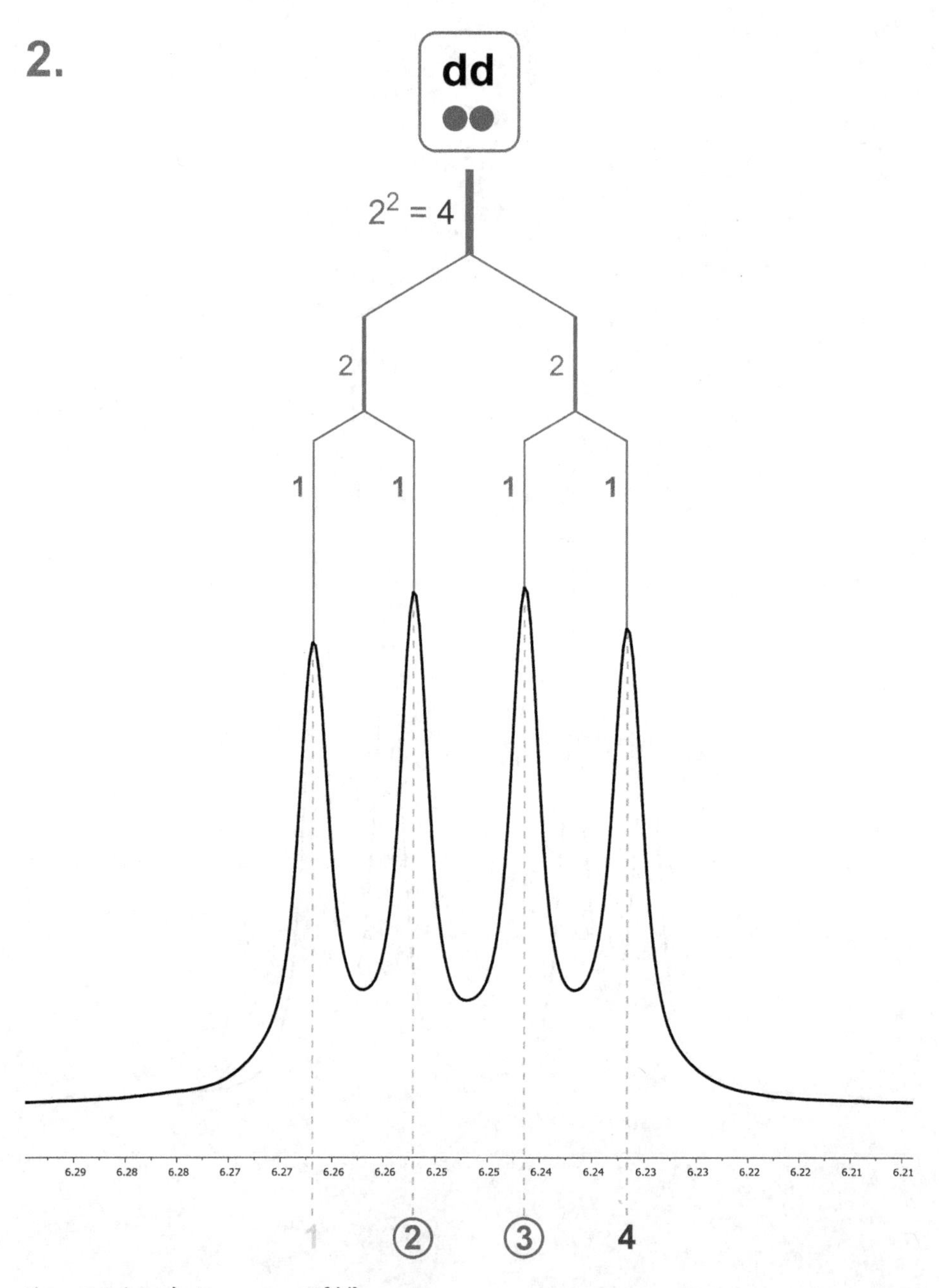

Figure 5.7: Introductory answer 2 [dd].

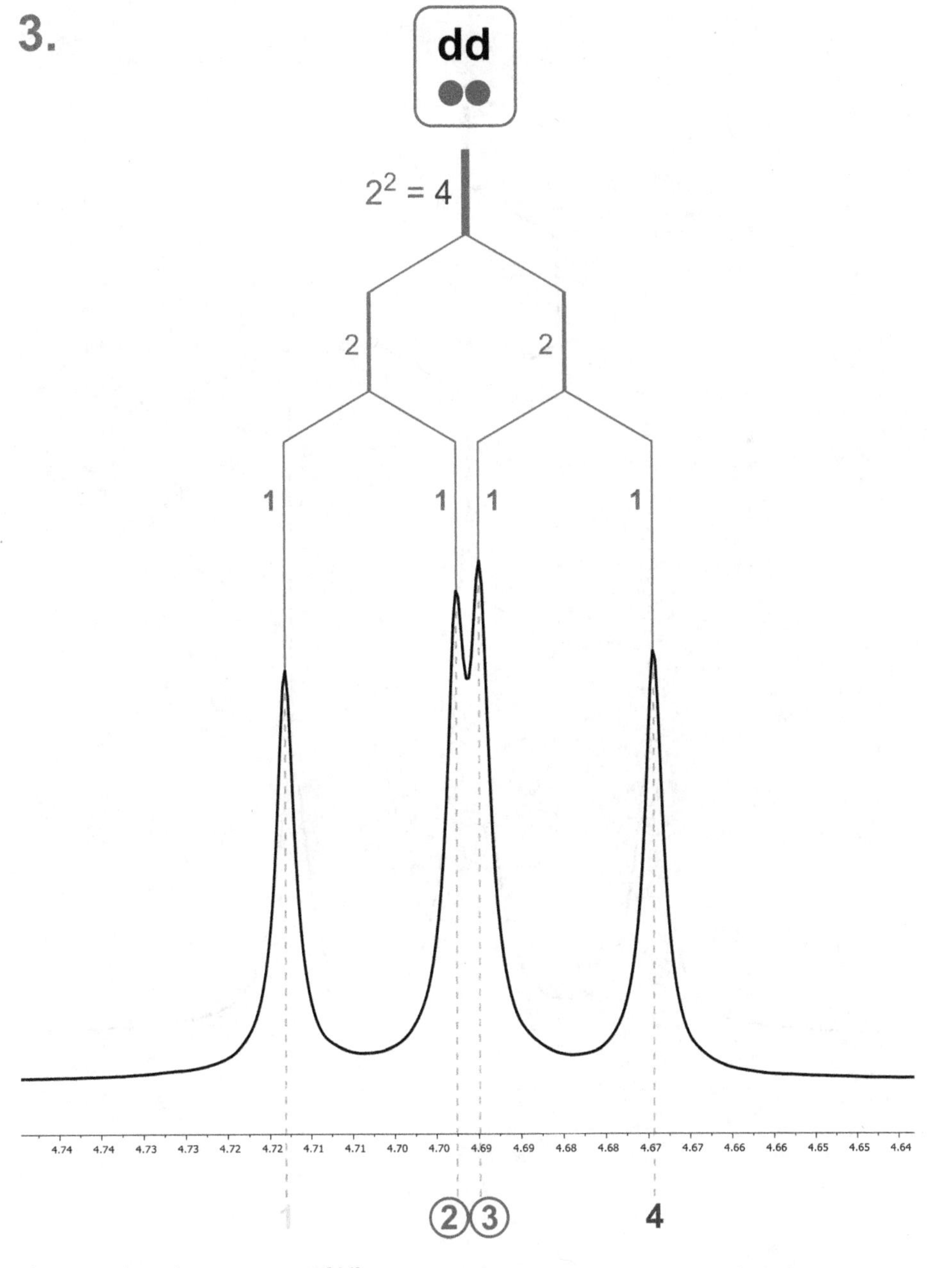

Figure 5.8: Introductory answer 3 [dd].

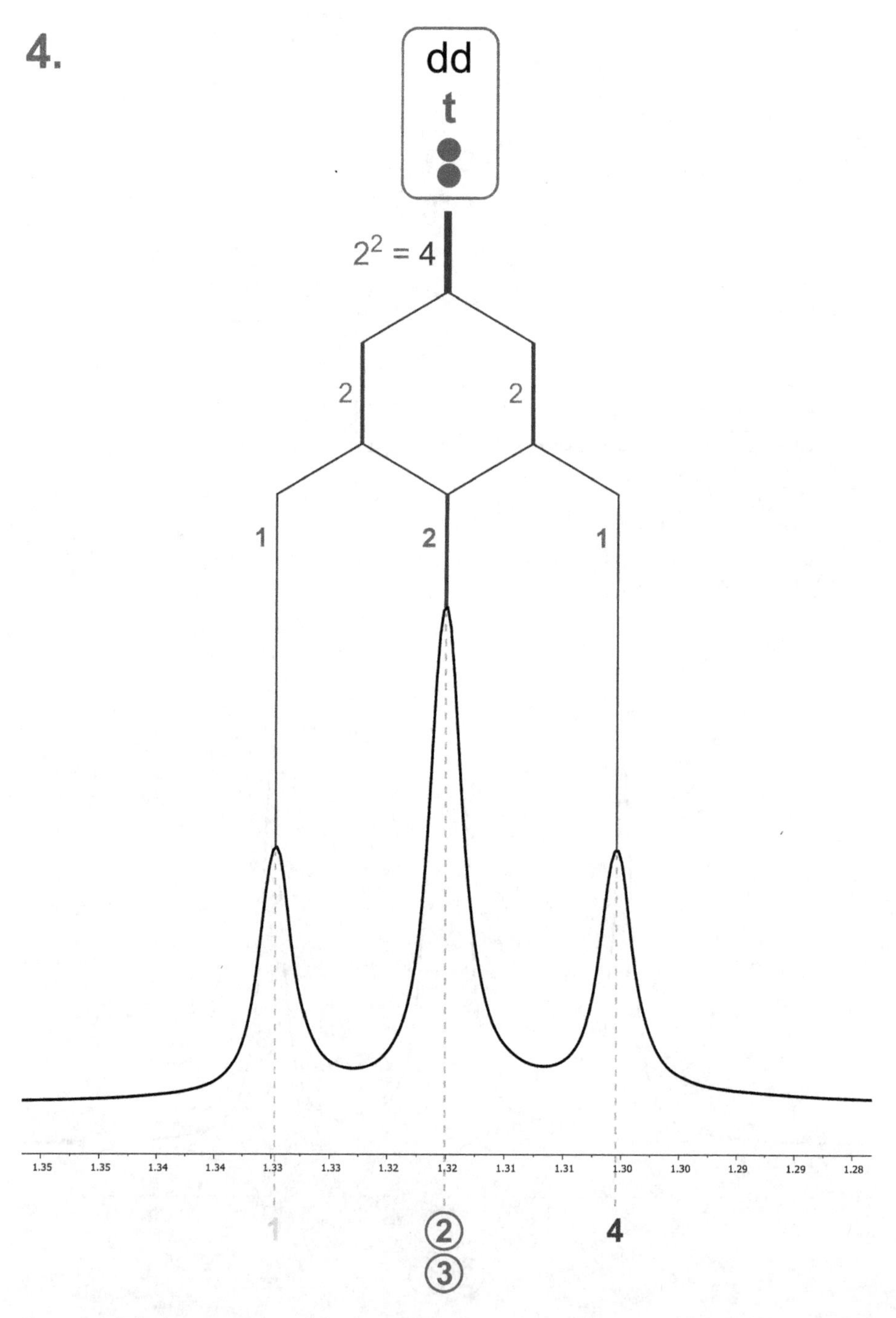

Figure 5.9: Introductory answer 4 [t].

5.2 Intermediate Level: doublet of doublet of doublets [ddd]

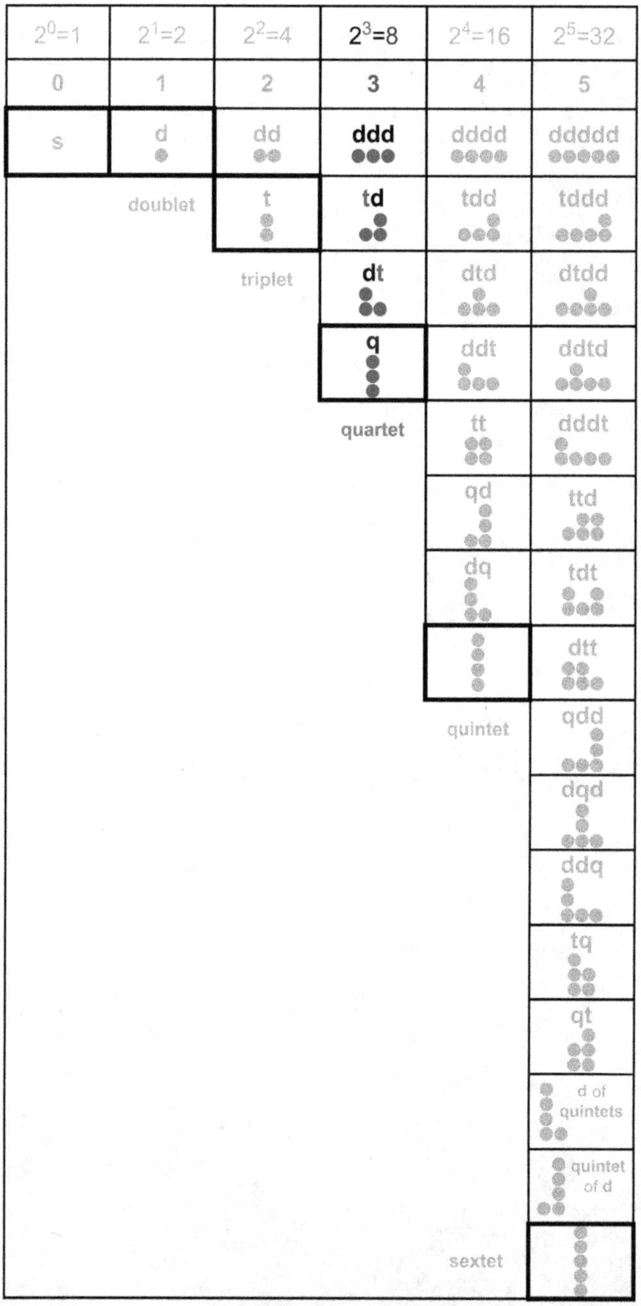

$2^0=1$	$2^1=2$	$2^2=4$	$2^3=8$	$2^4=16$	$2^5=32$
0	1	2	3	4	5
s	d	dd	ddd	dddd	ddddd
	doublet	t	td	tdd	tddd
		triplet	dt	dtd	dtdd
			q	ddt	ddtd
			quartet	tt	dddt
				qd	ttd
				dq	tdt
					dtt
				quintet	qdd
					dqd
					ddq
					tq
					qt
					d of quintets
					quintet of d
				sextet	

Figure 5.10: Multiplets with three J-coupling constants ($n = 3$: J_1, J_2, J_3).

5.2.1 Intermediate Exercises

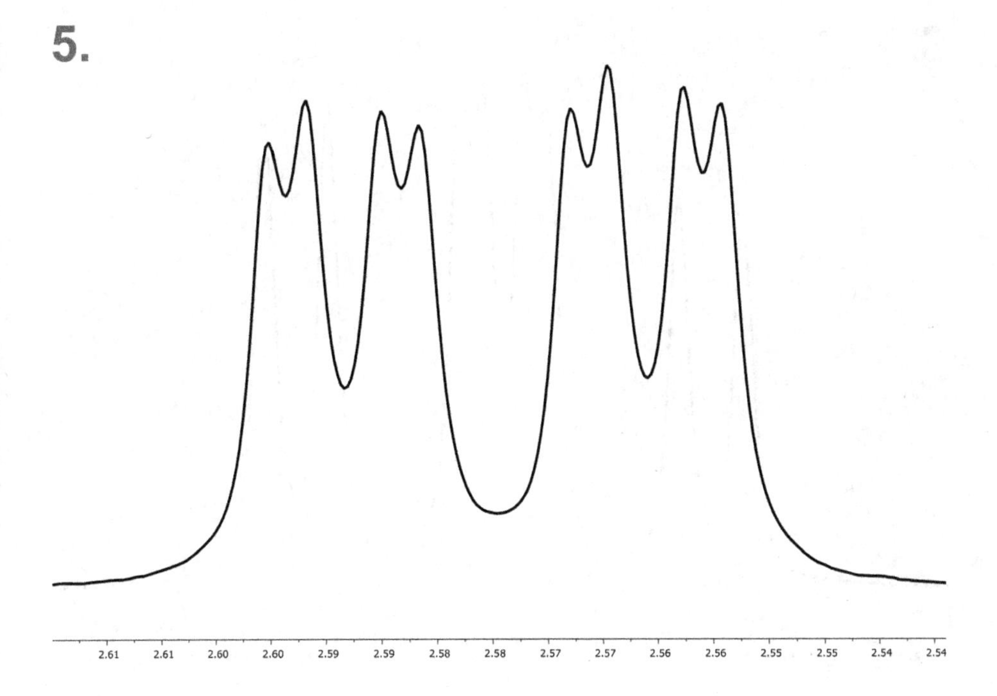

Figure 5.11: Intermediate exercise 5.

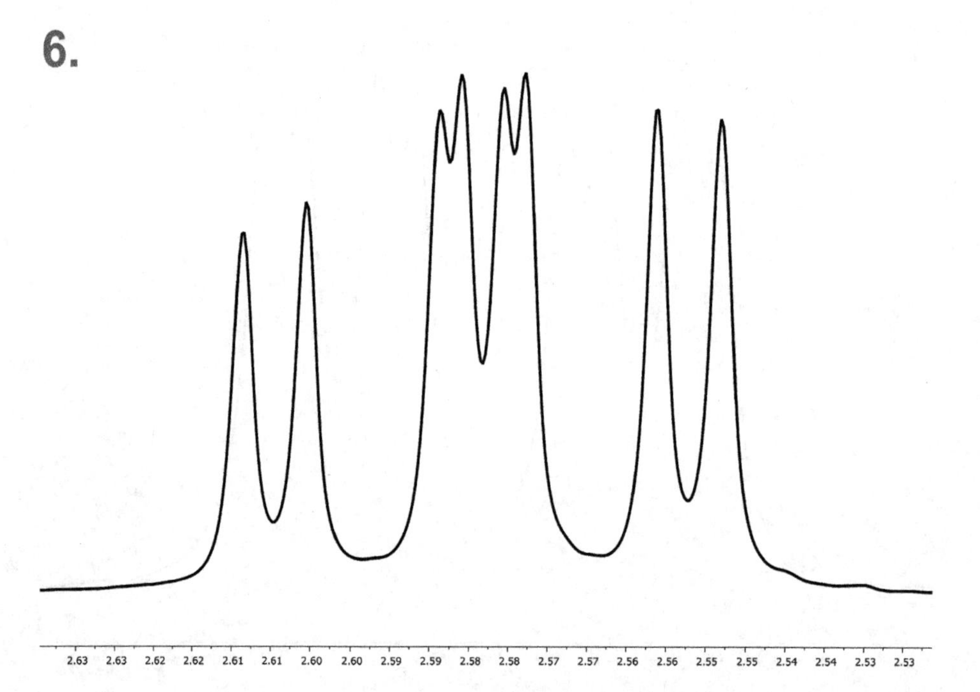

Figure 5.12: Intermediate exercise 6.

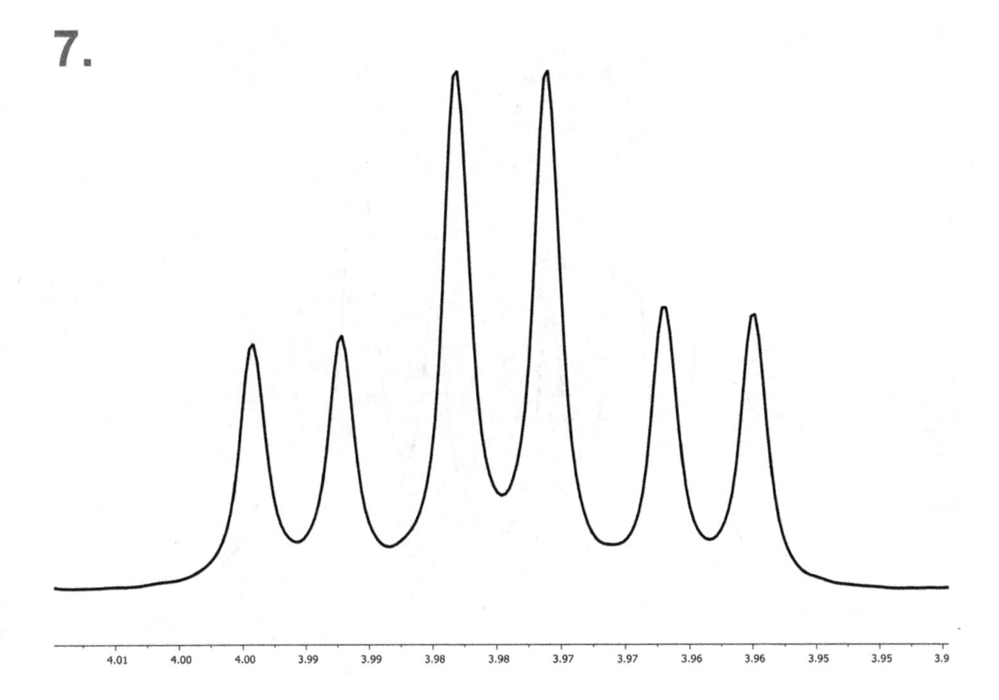

Figure 5.13: Intermediate exercise 7.

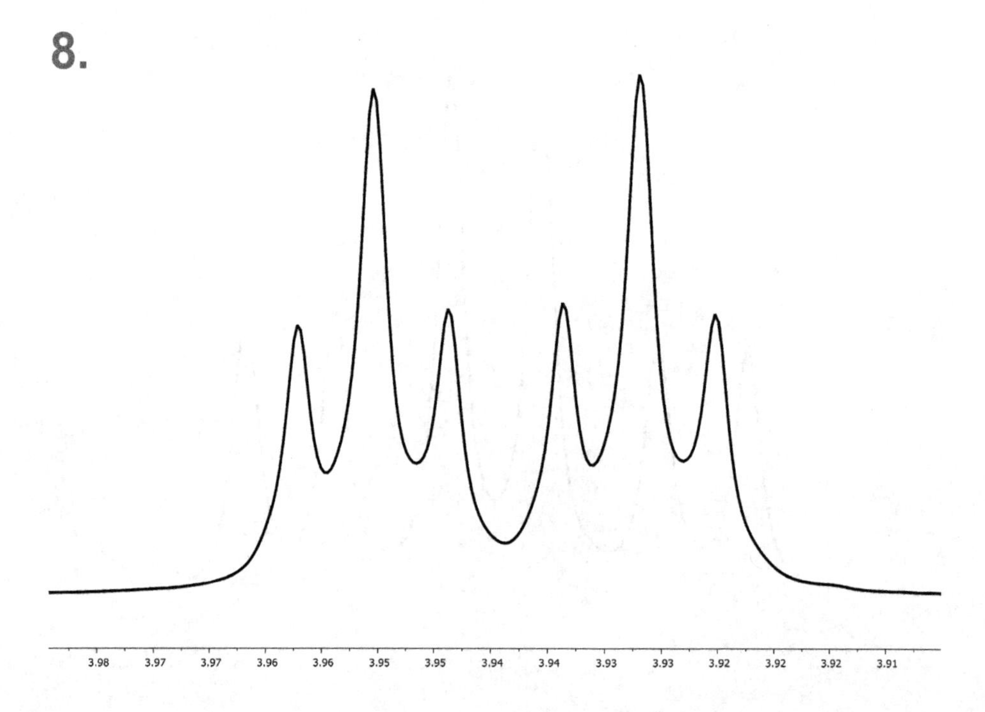

Figure 5.14: Intermediate exercise 8.

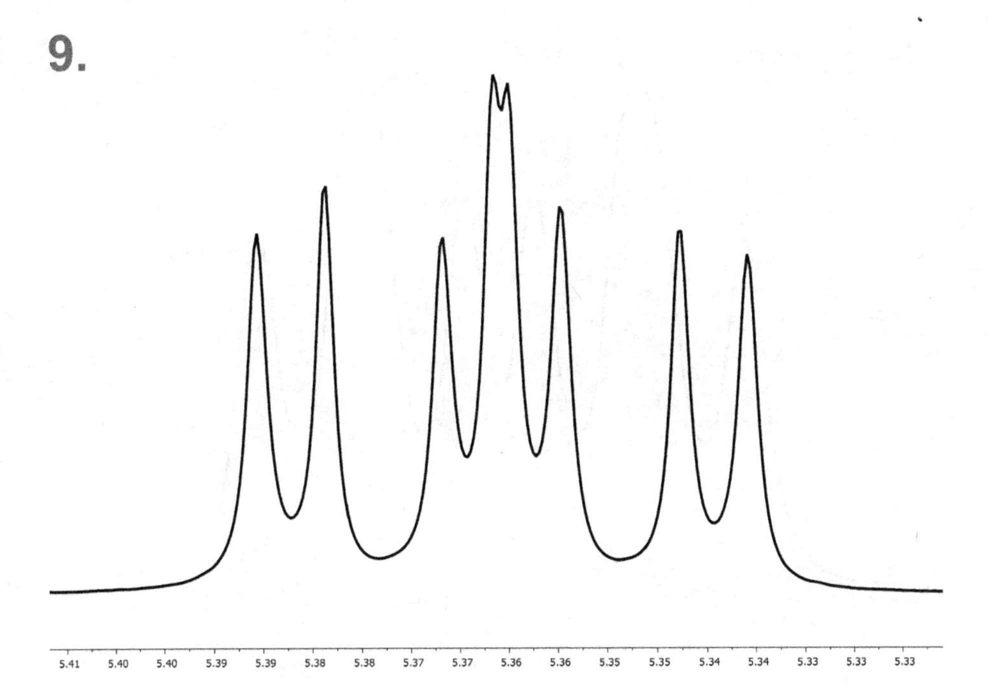

Figure 5.15: Intermediate exercise 9.

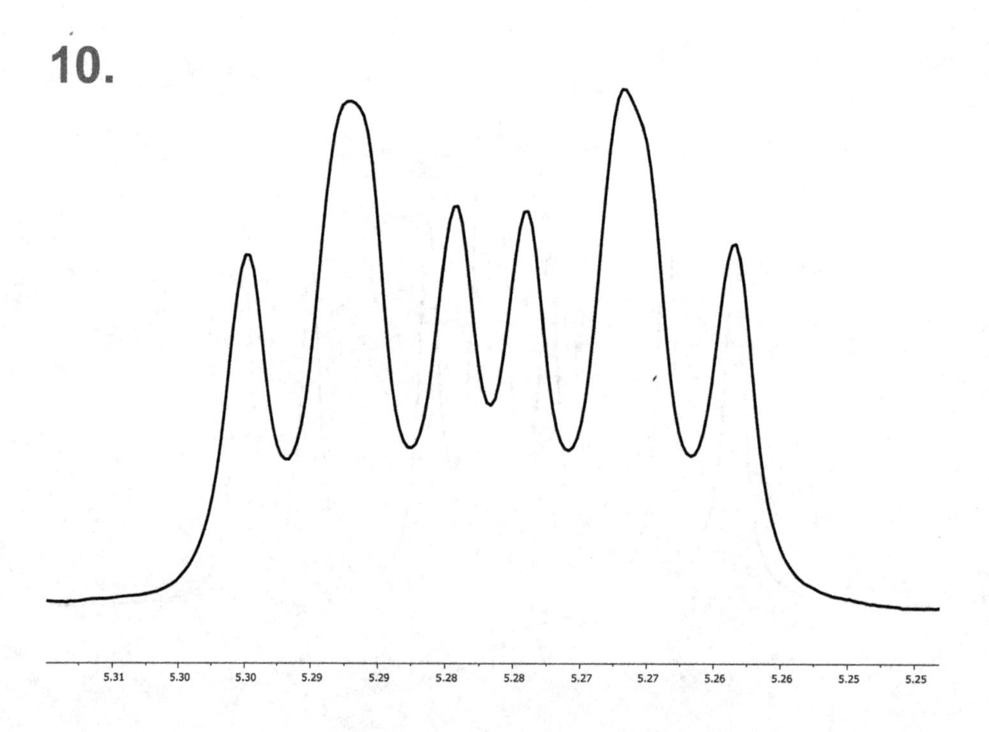

Figure 5.16: Intermediate exercise 10.

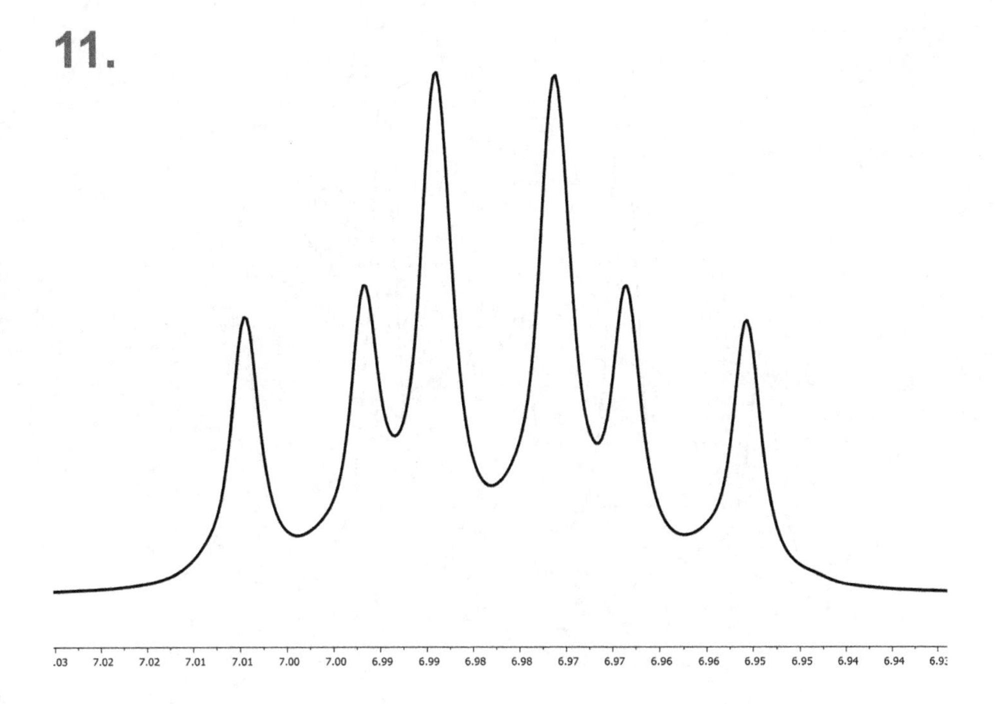

Figure 5.17: Intermediate exercise 11.

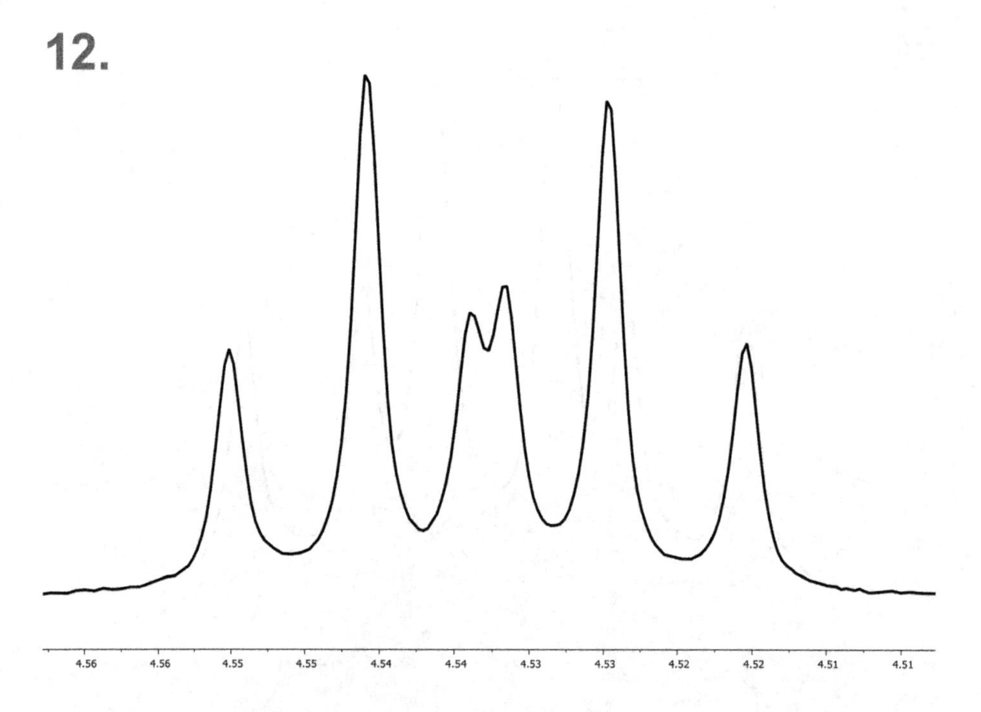

Figure 5.18: Intermediate exercise 12.

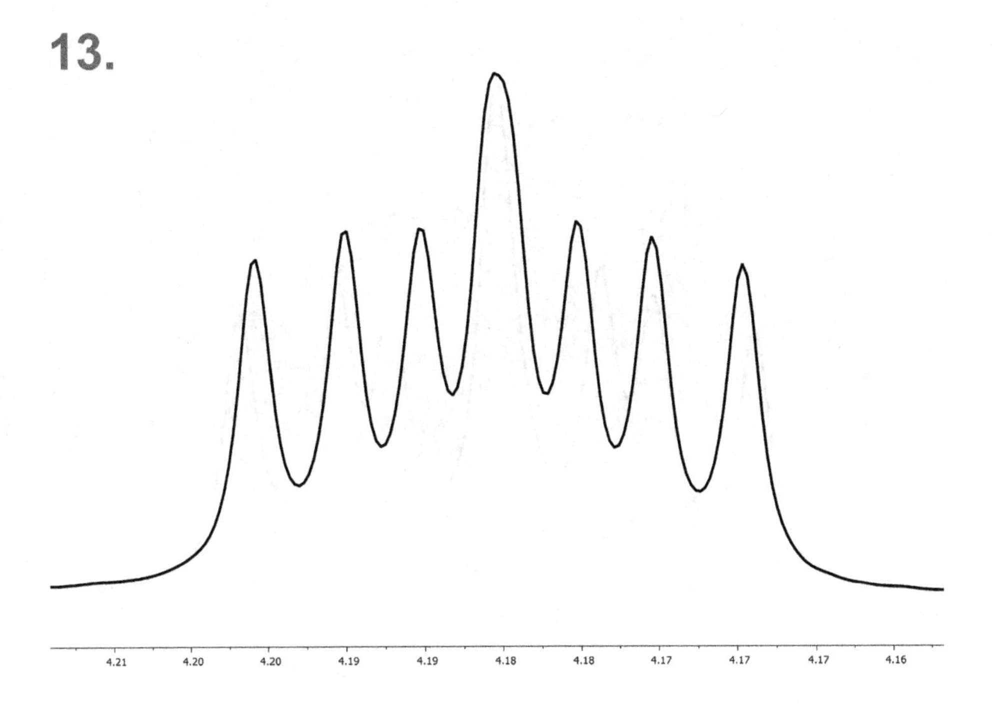

Figure 5.19: Intermediate exercise 13.

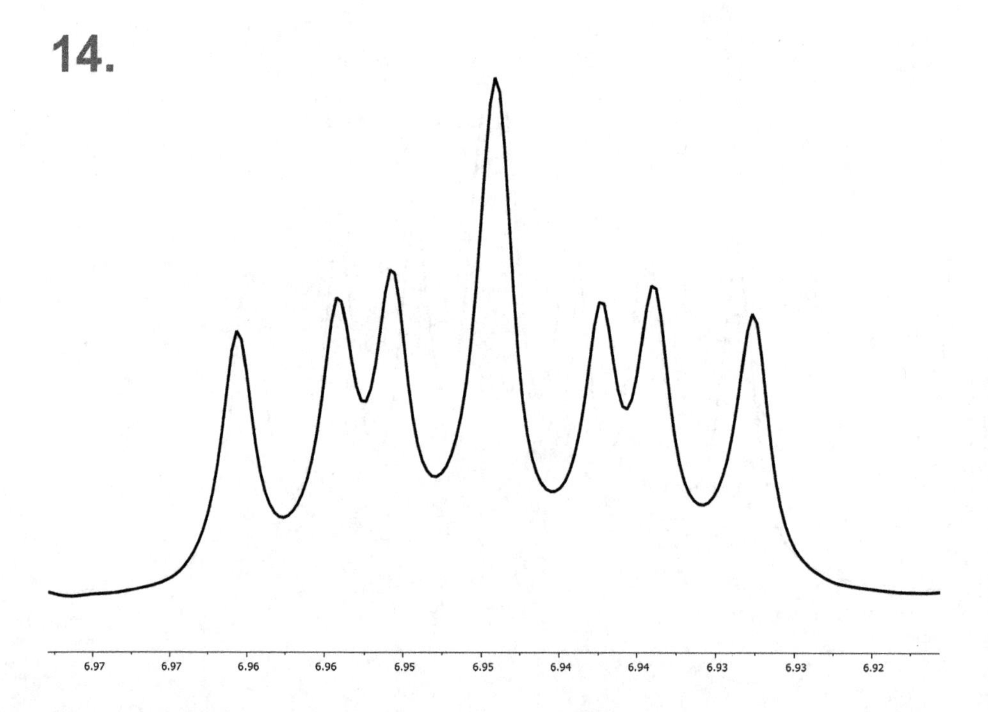

Figure 5.20: Intermediate exercise 14.

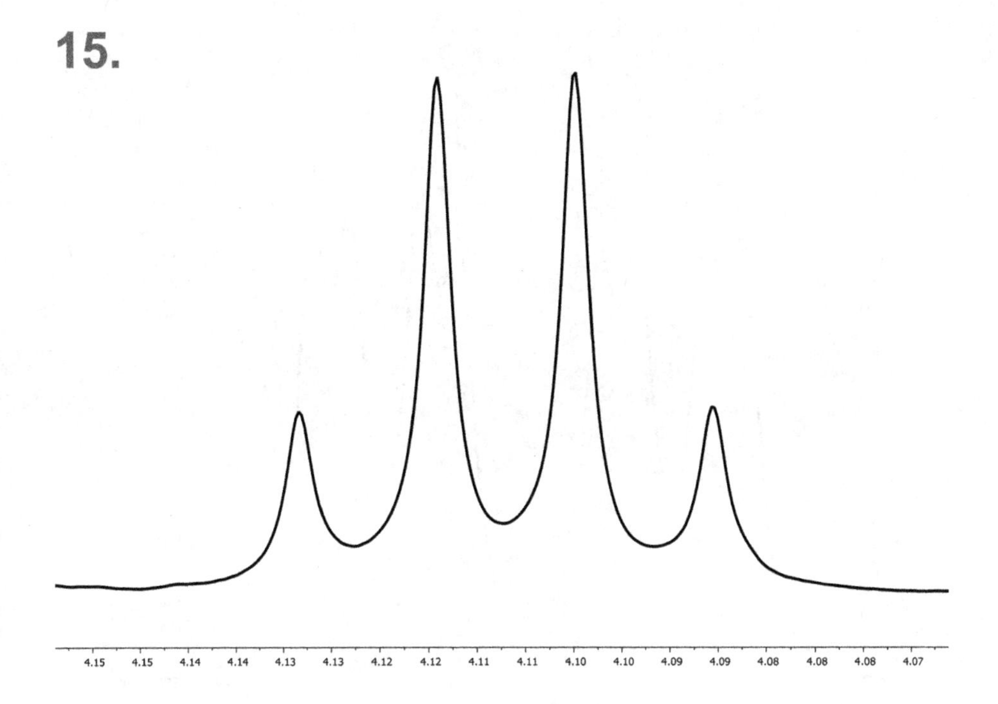

Figure 5.21: Intermediate exercise 15.

5.2.2 Intermediate Answers

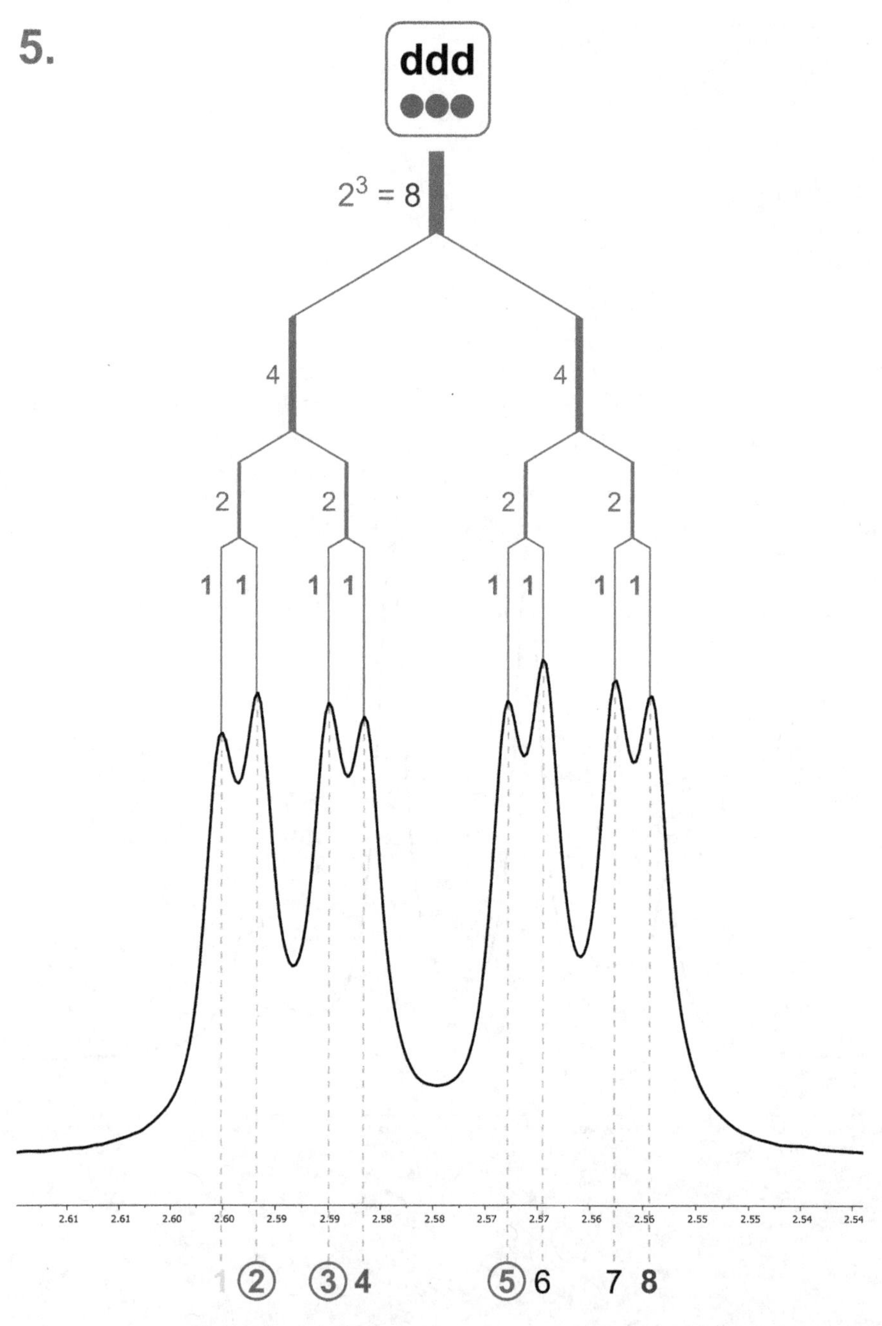

Figure 5.22: Intermediate answer 5 [ddd].

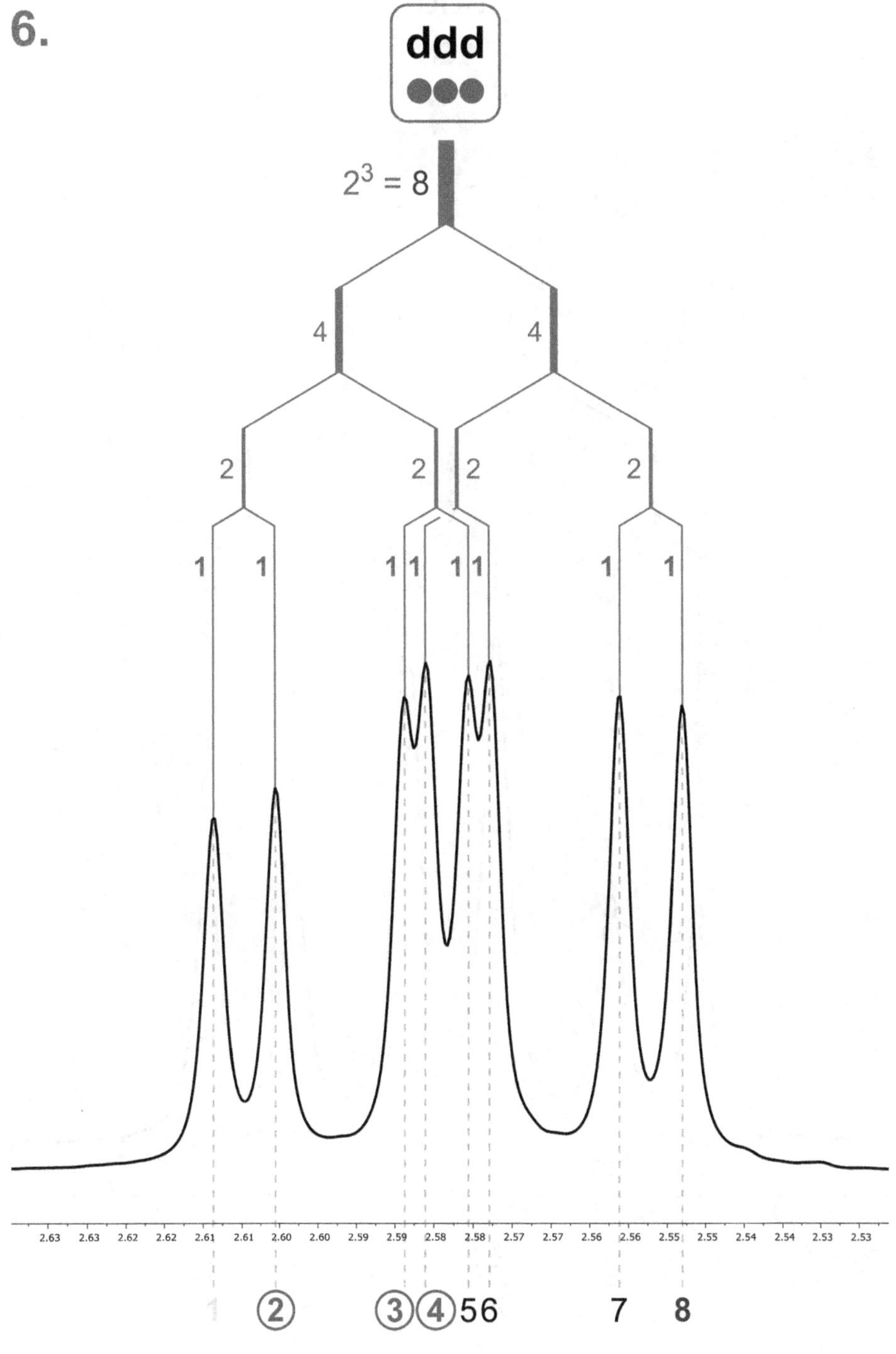

Figure 5.23: Intermediate answer 6 [ddd].

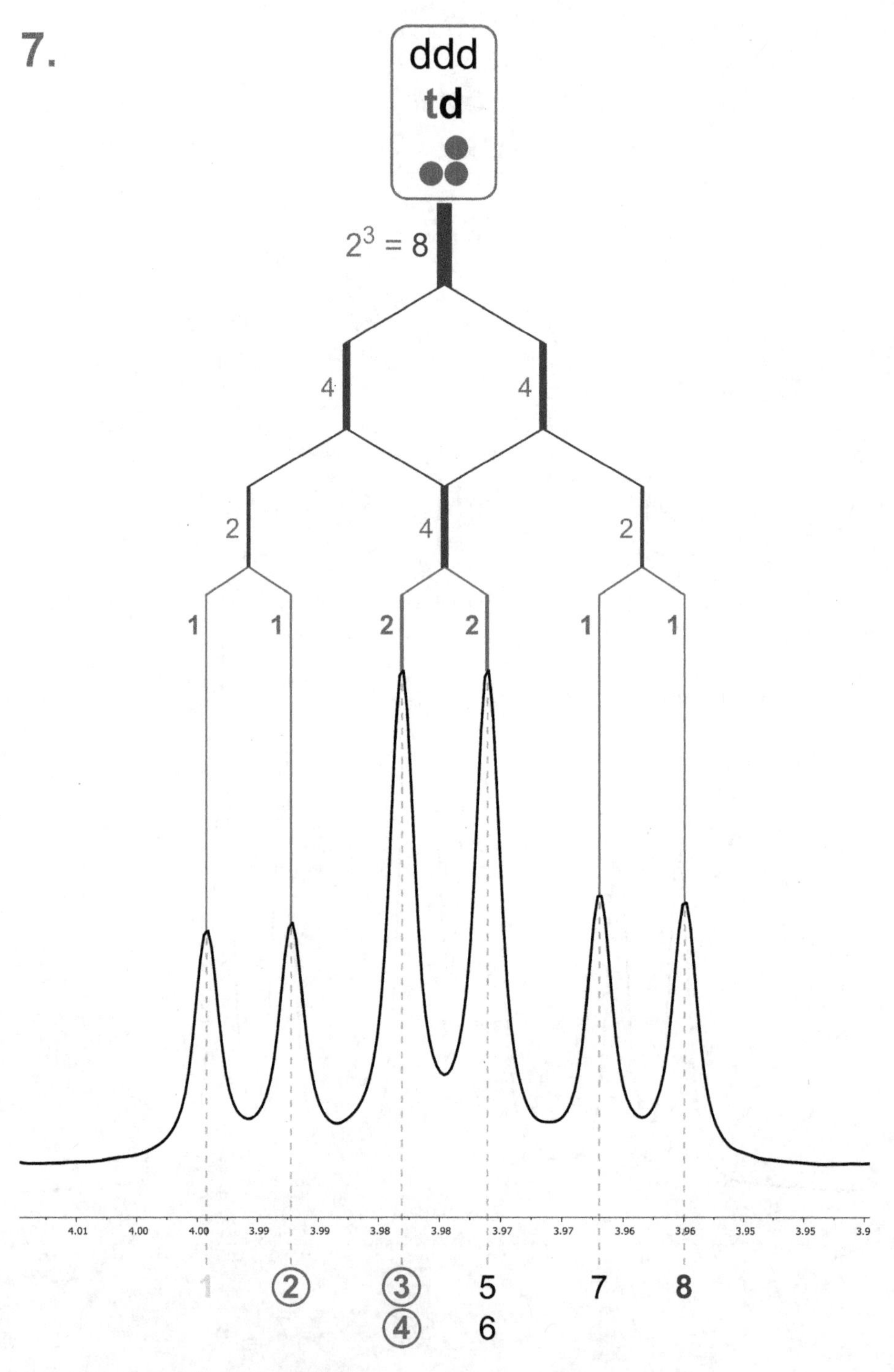

Figure 5.24: Intermediate answer 7 [td].

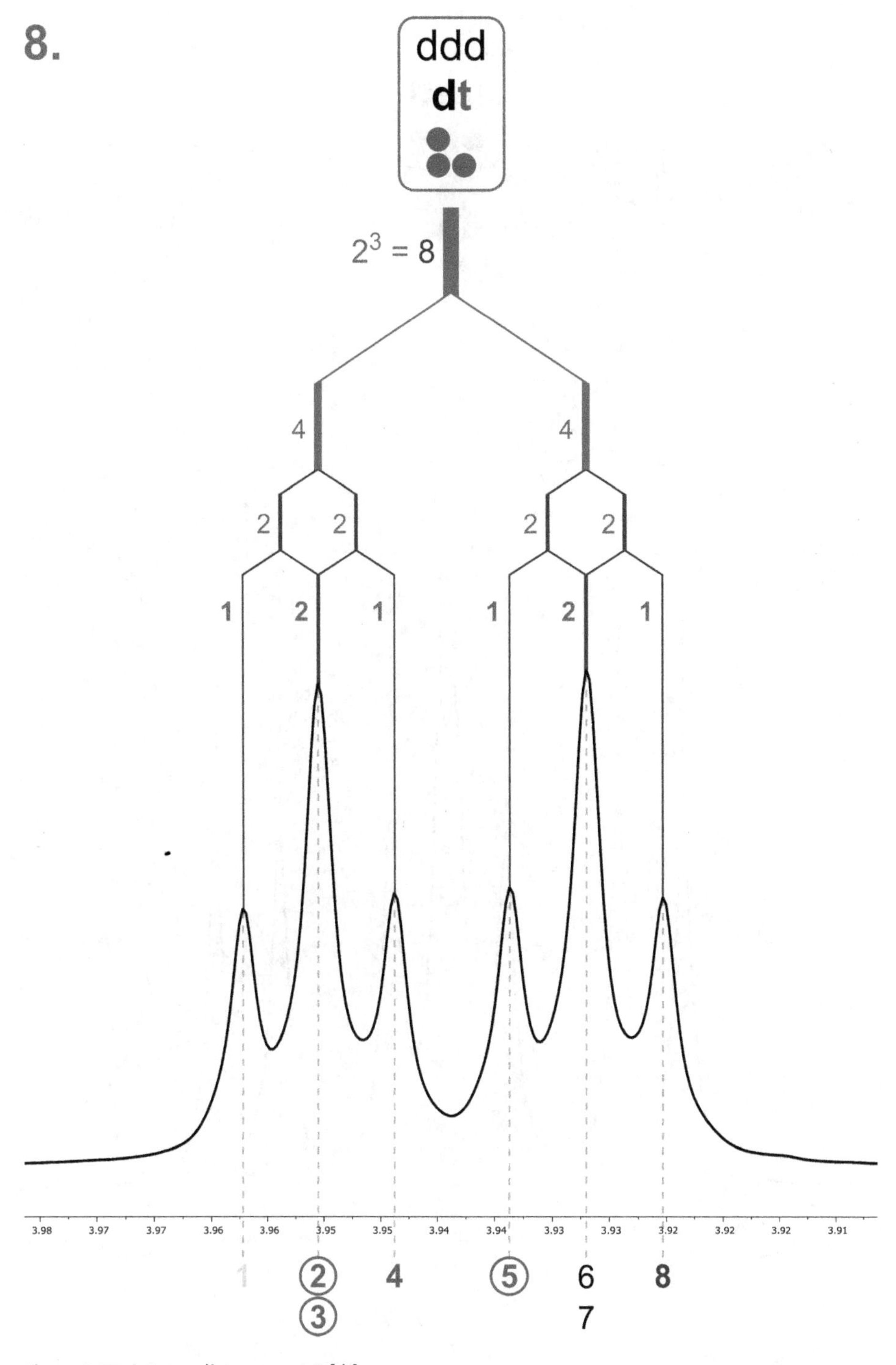

Figure 5.25: Intermediate answer 8 [dt].

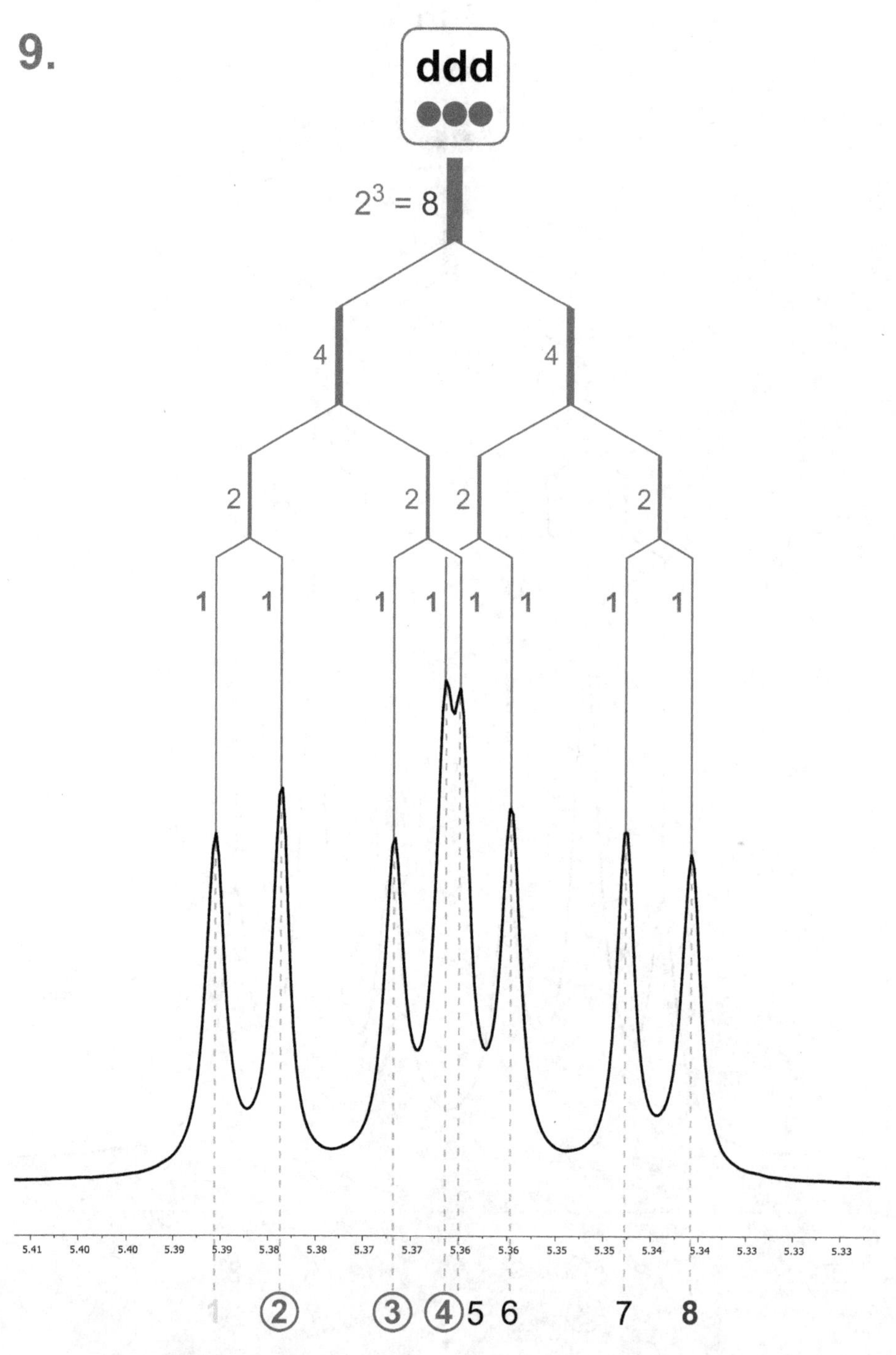

Figure 5.26: Intermediate answer 9 [ddd].

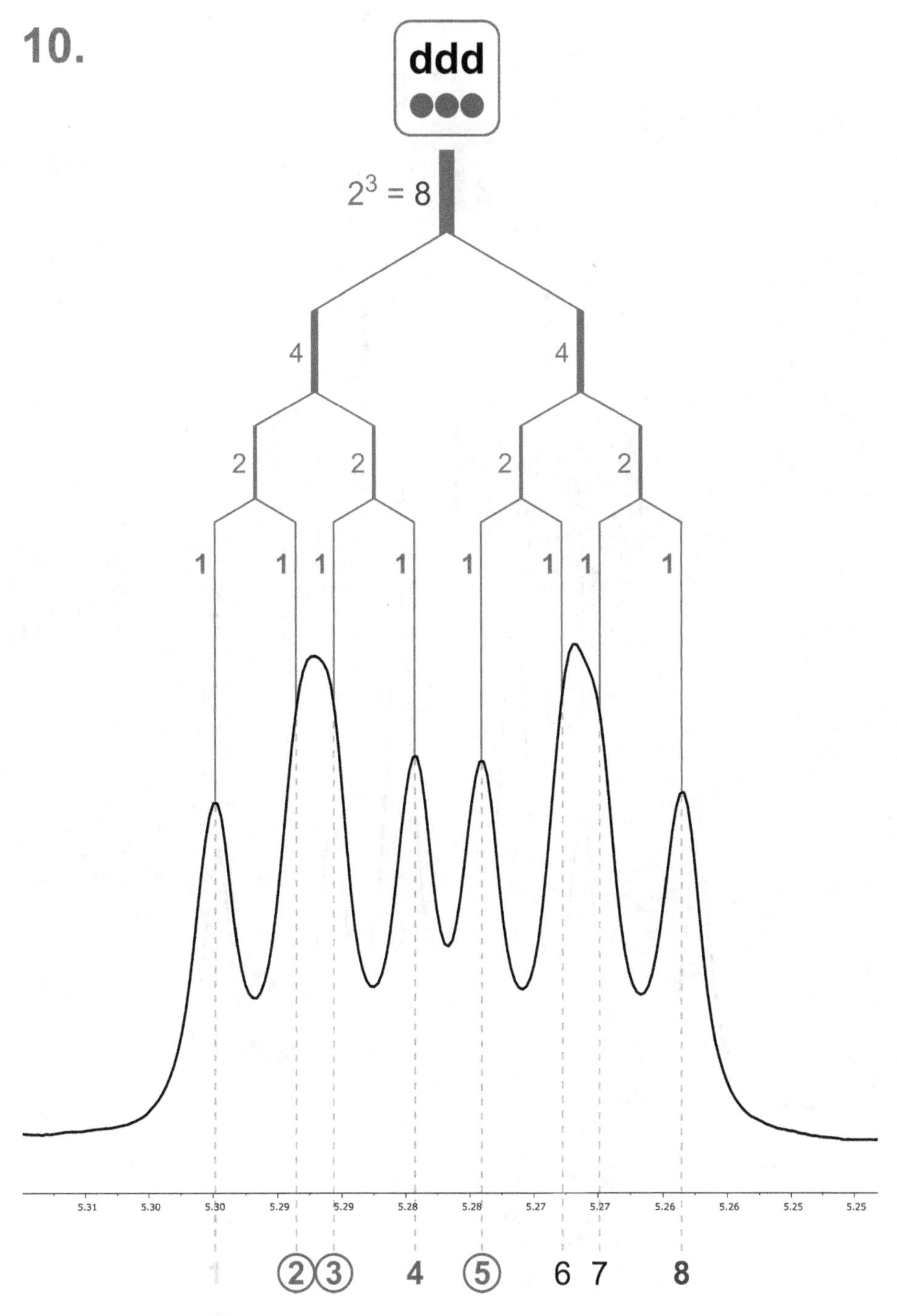

Figure 5.27: Intermediate answer 10 [ddd].

11.

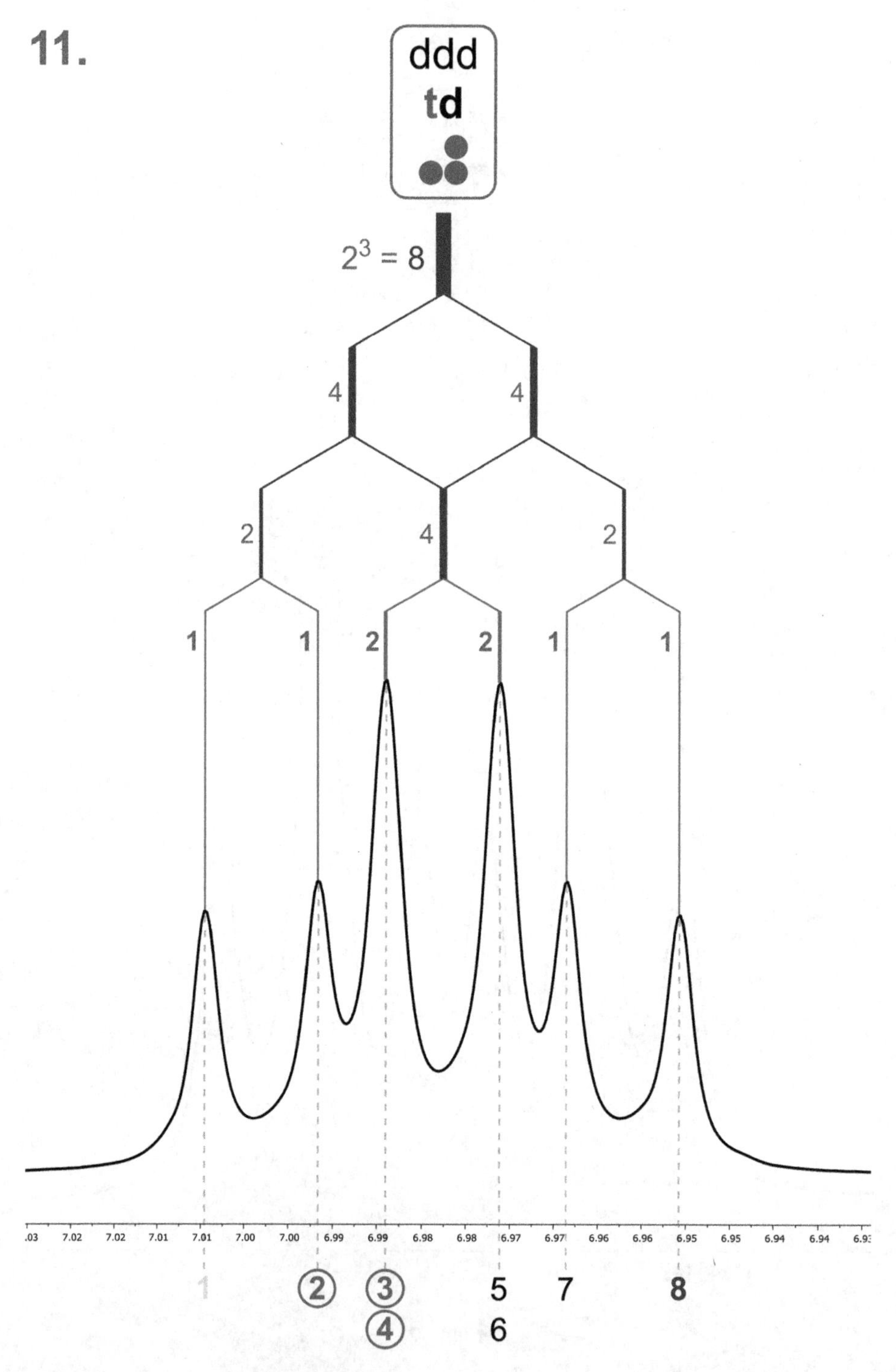

Figure 5.28: Intermediate answer 11 [td].

12.

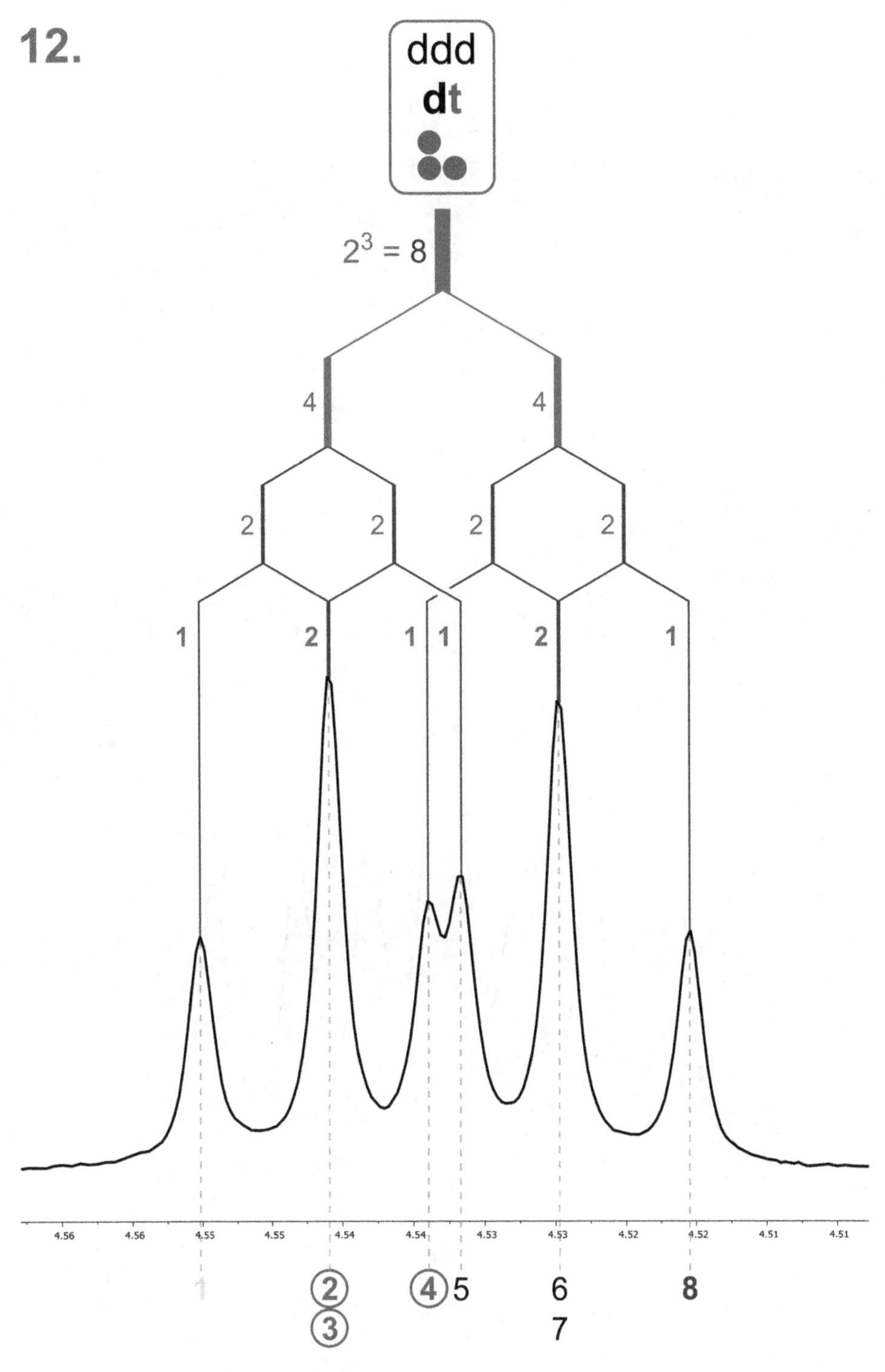

Figure 5.29: Intermediate answer 12 [dt].

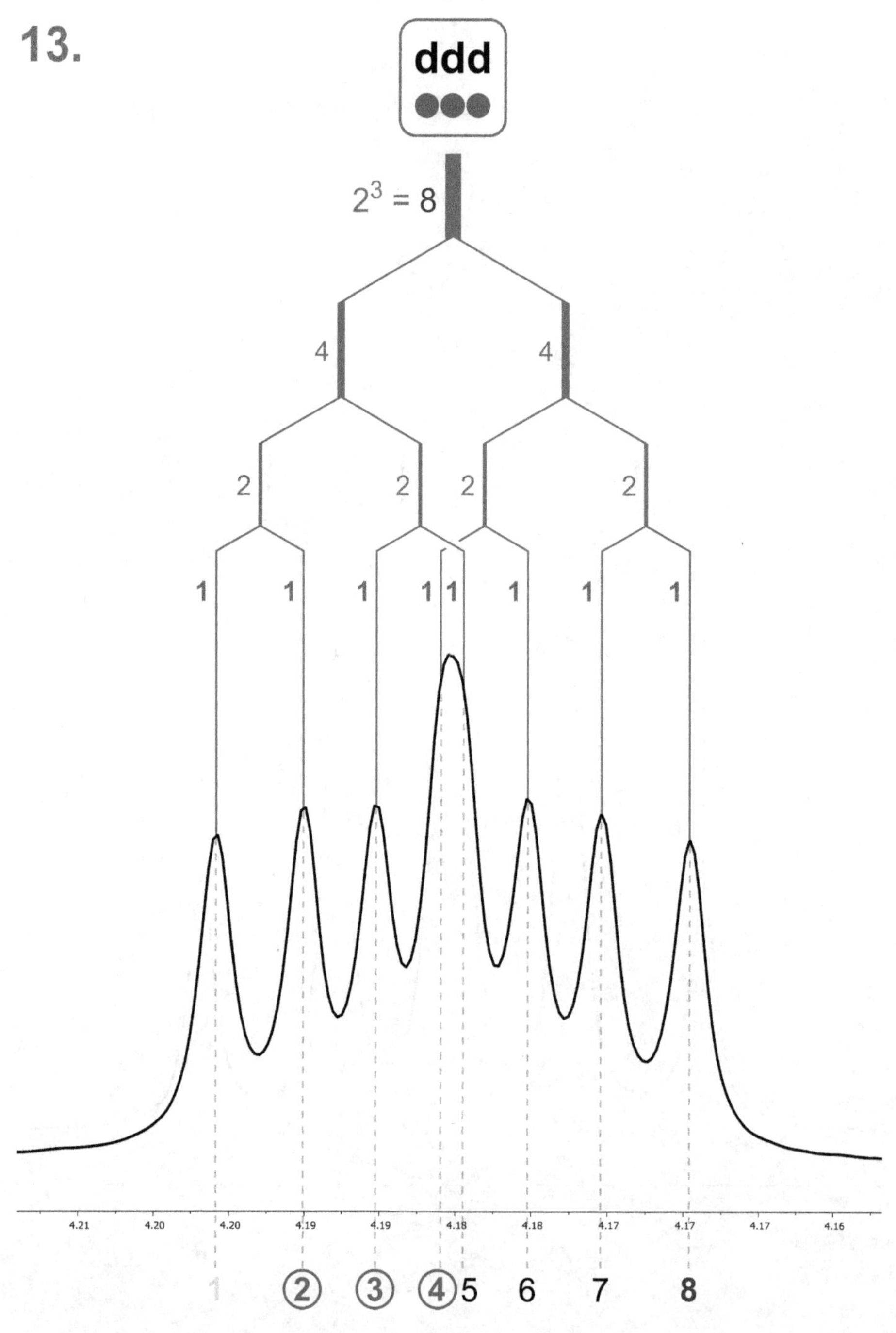

Figure 5.30: Intermediate answer 13 [ddd].

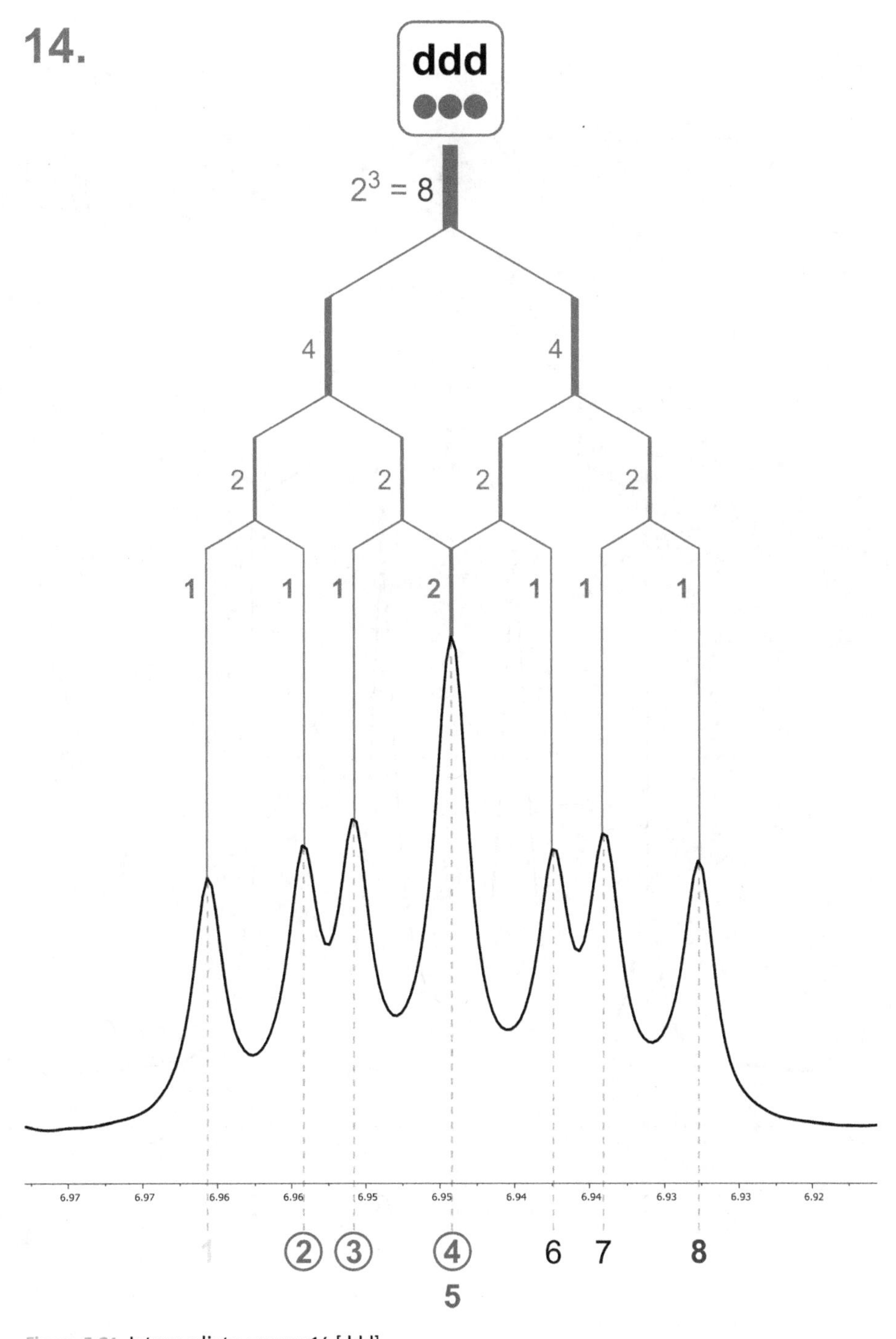

Figure 5.31: Intermediate answer 14 [ddd].

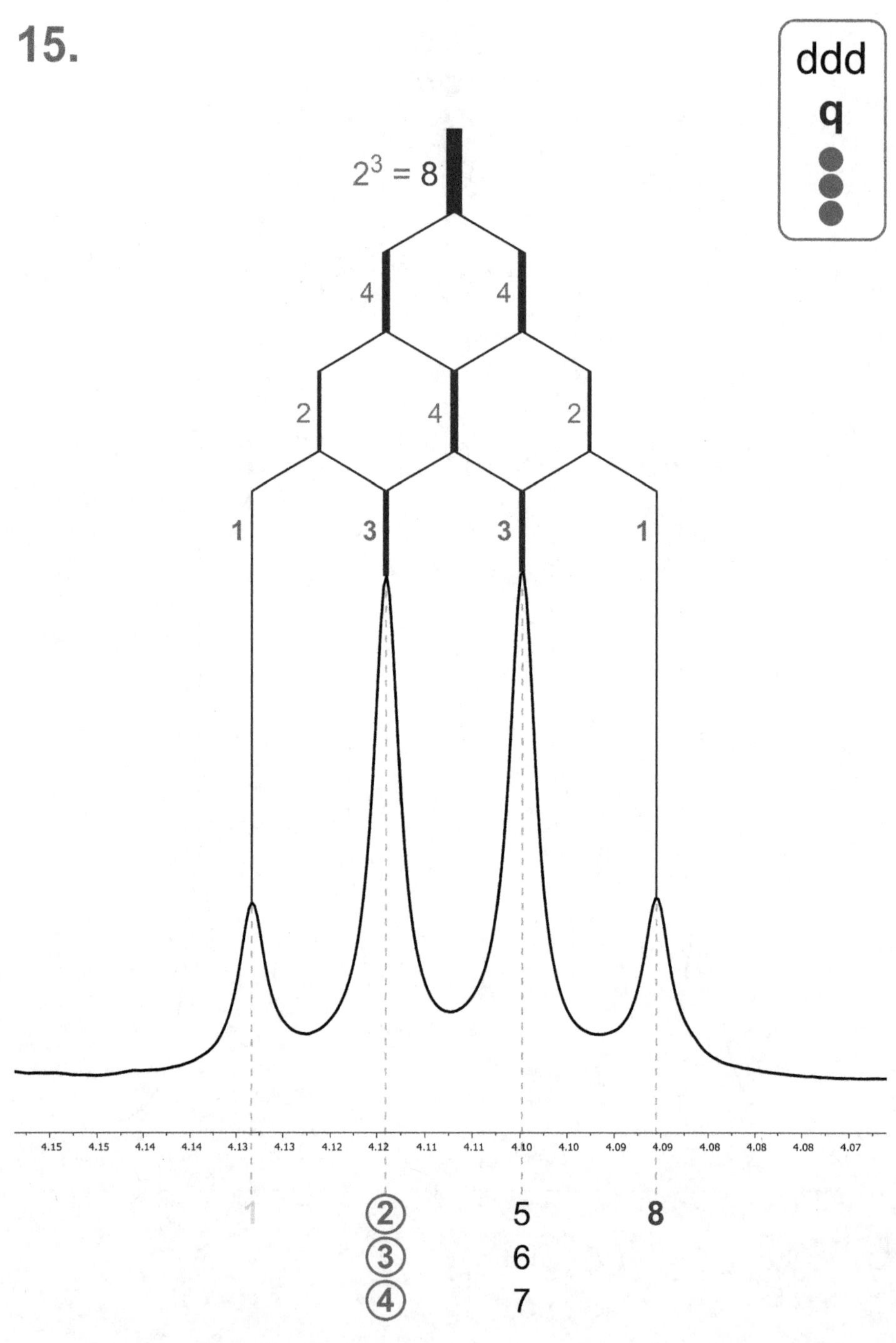

Figure 5.32: Intermediate answer 15 [q].

5.3 Advanced Level: doublet of doublet of doublet of doublets [dddd]

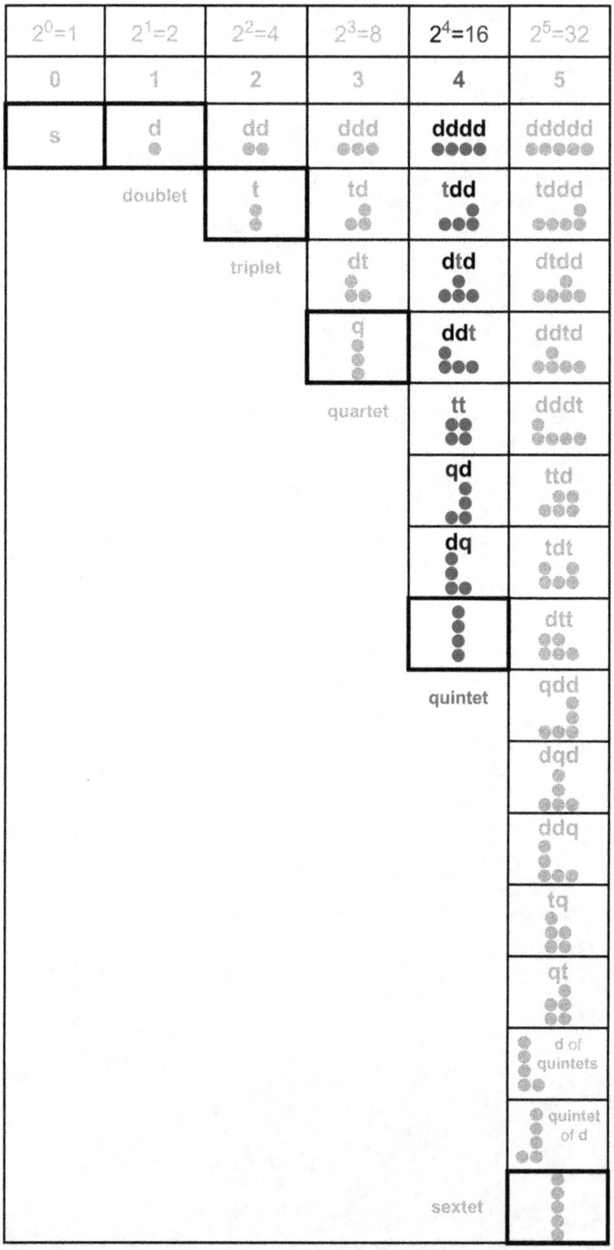

$2^0=1$	$2^1=2$	$2^2=4$	$2^3=8$	$2^4=16$	$2^5=32$
0	1	2	3	4	5
s	d	dd	ddd	**dddd**	ddddd
	doublet	t	td	**tdd**	tddd
		triplet	dt	**dtd**	dtdd
			q	**ddt**	ddtd
			quartet	**tt**	dddt
				qd	ttd
				dq	tdt
					dtt
				quintet	qdd
					dqd
					ddq
					tq
					qt
					d of quintets
					quintet of d
				sextet	

Figure 5.33: Multiplets with four J-coupling constants ($n = 4$: J_1, J_2, J_3, J_4).

5.3.1 Advanced Exercises

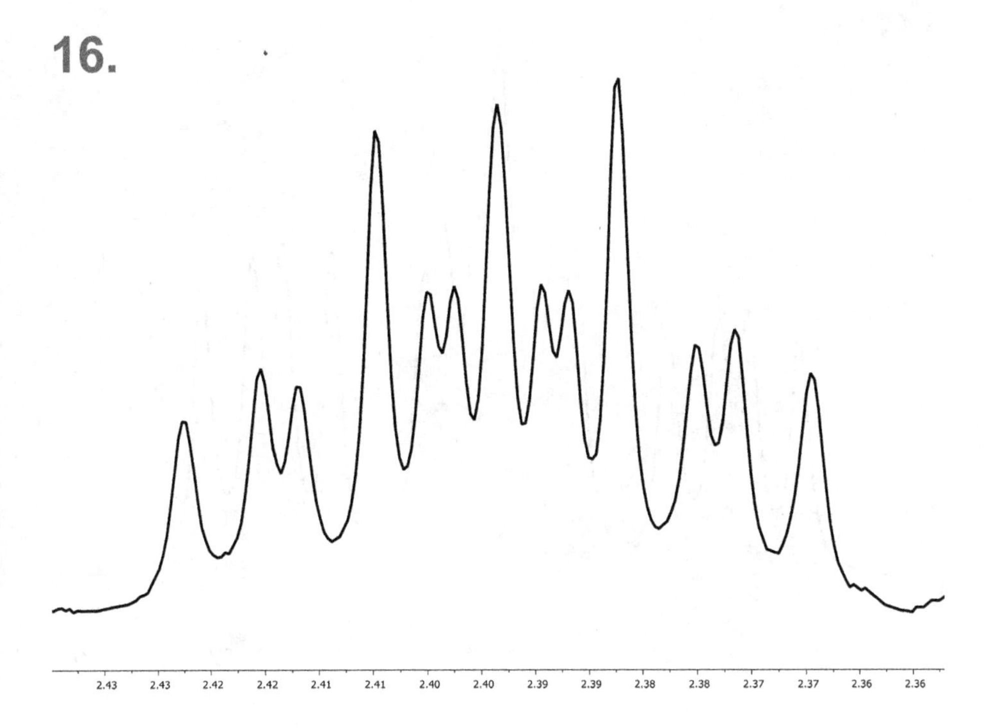

Figure 5.34: Advanced exercise 16.

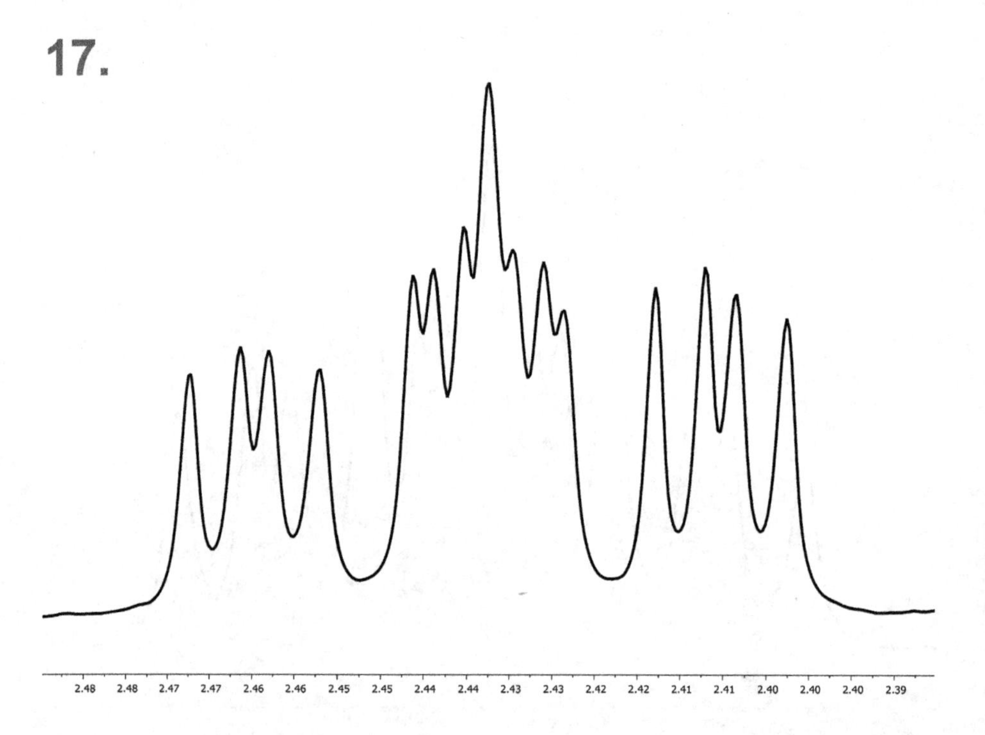

Figure 5.35: Advanced exercise 17.

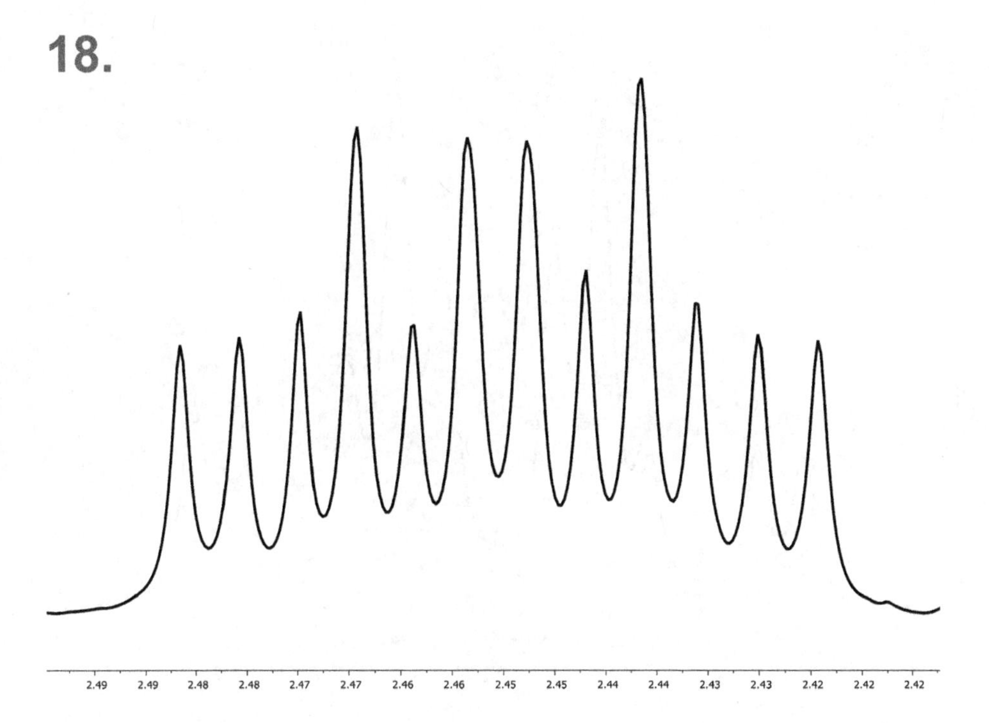

Figure 5.36: Advanced exercise 18.

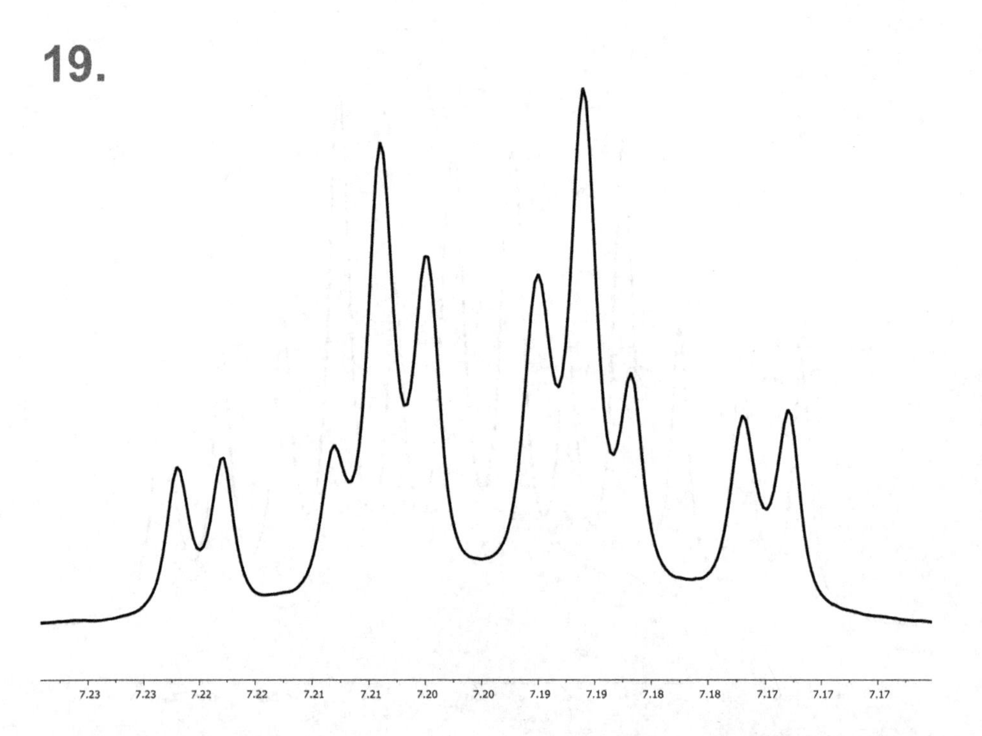

Figure 5.37: Advanced exercise 19.

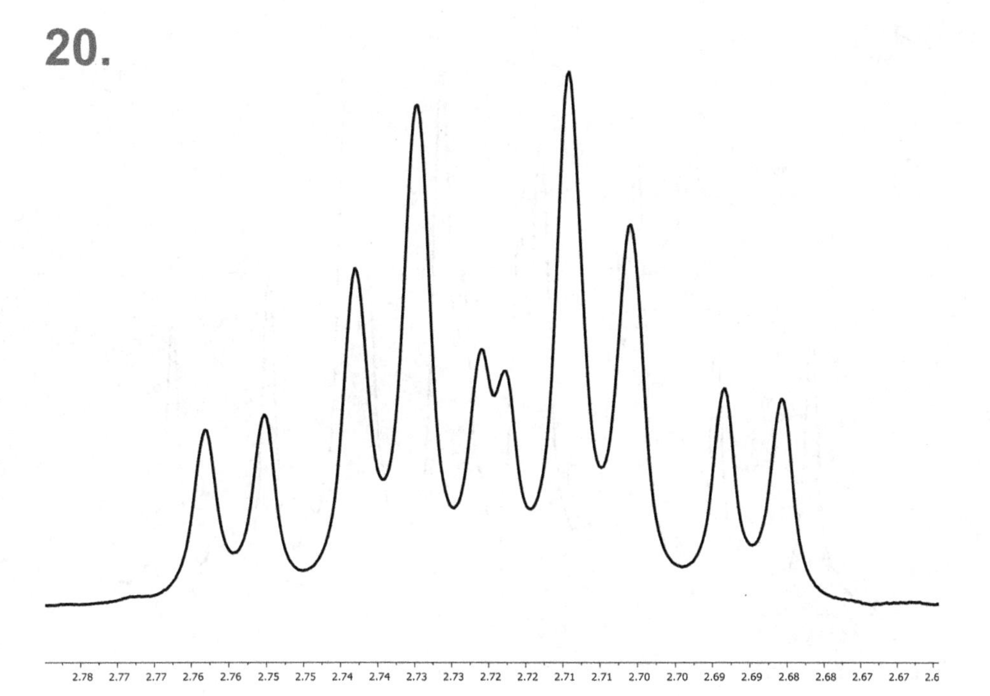

Figure 5.38: Advanced exercise 20.

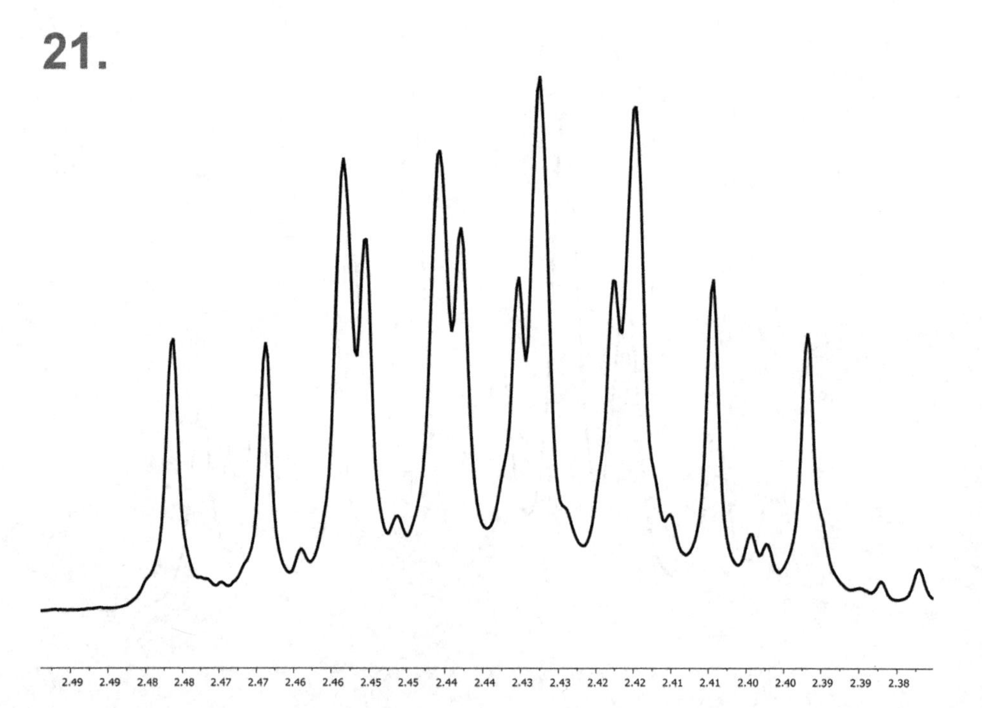

Figure 5.39: Advanced exercise 21.

22.

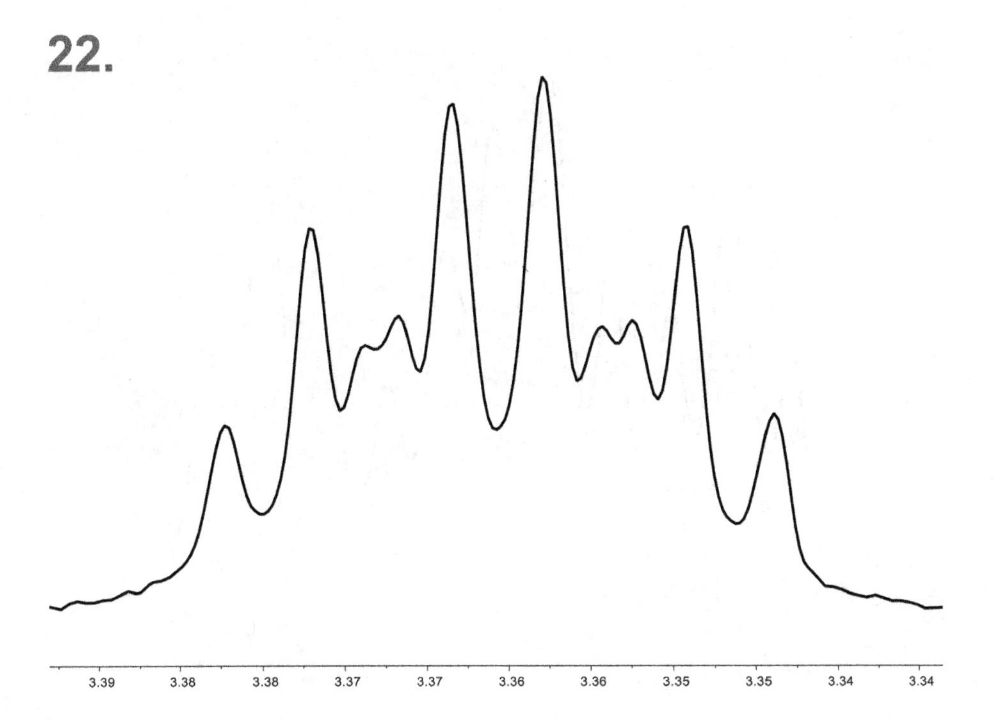

Figure 5.40: Advanced exercise 22.

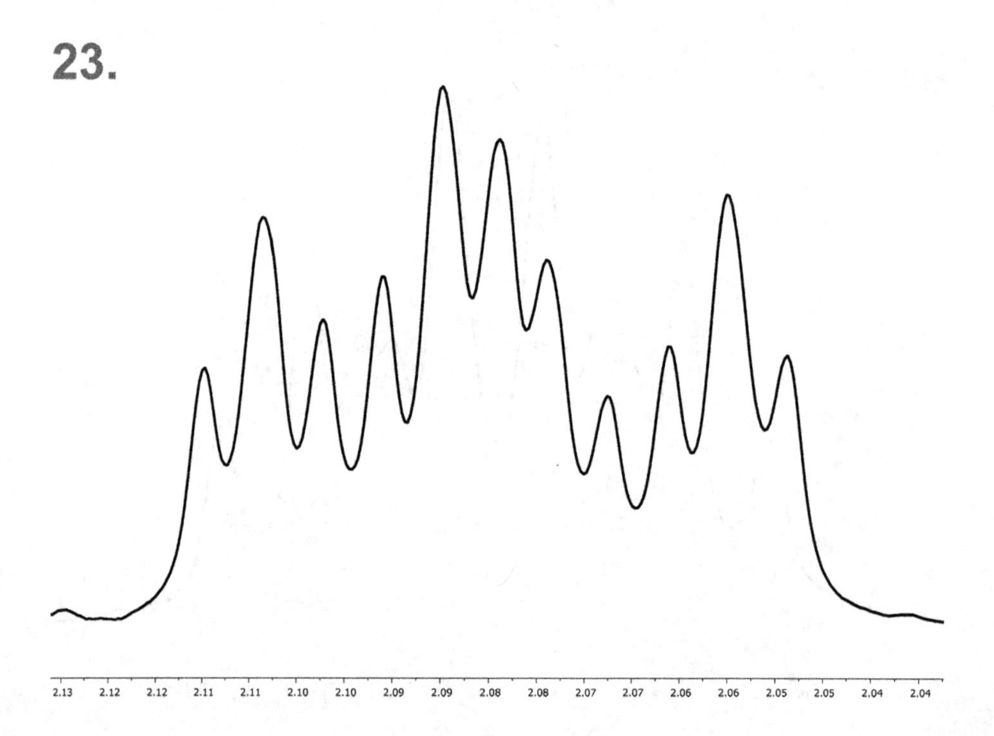

Figure 5.41: Advanced exercise 23.

24.

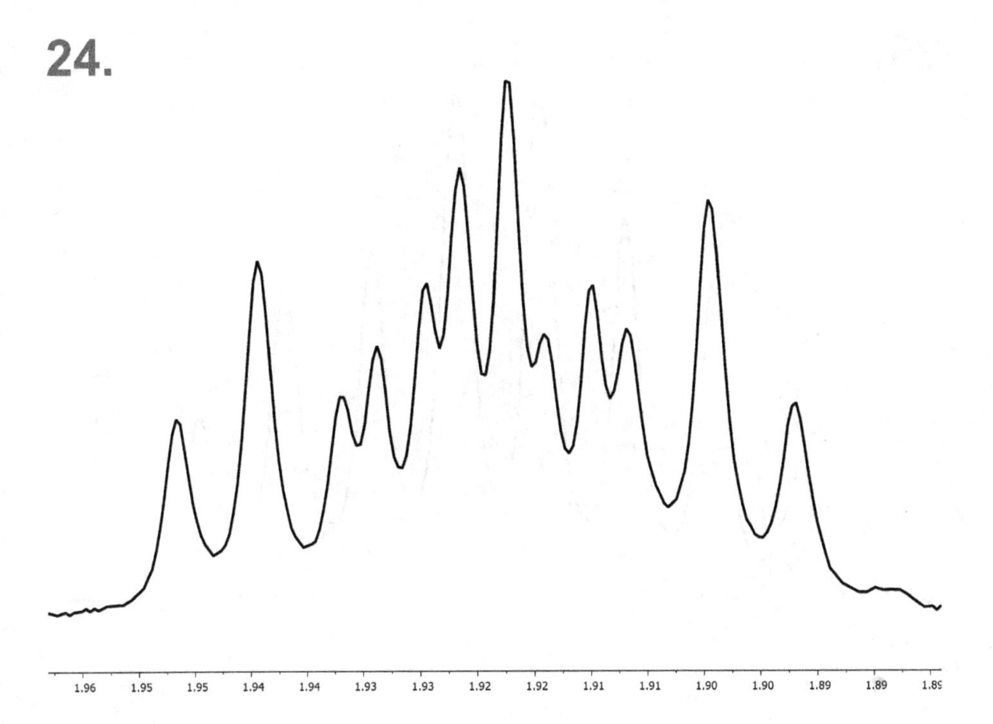

Figure 5.42: Advanced exercise 24.

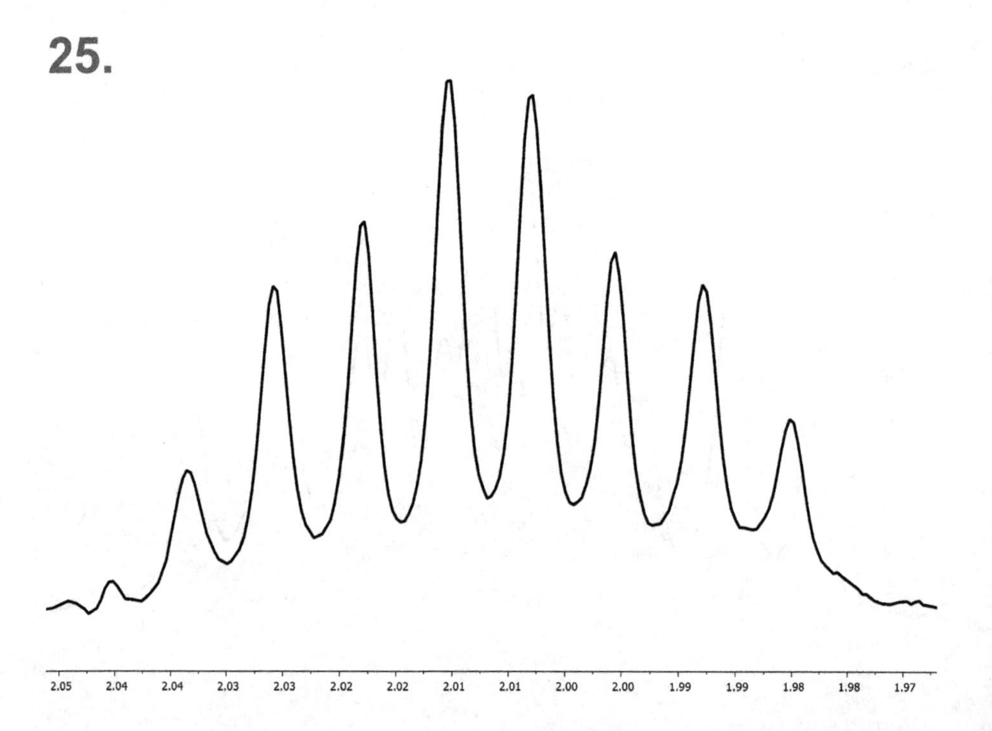

Figure 5.43: Advanced exercise 25.

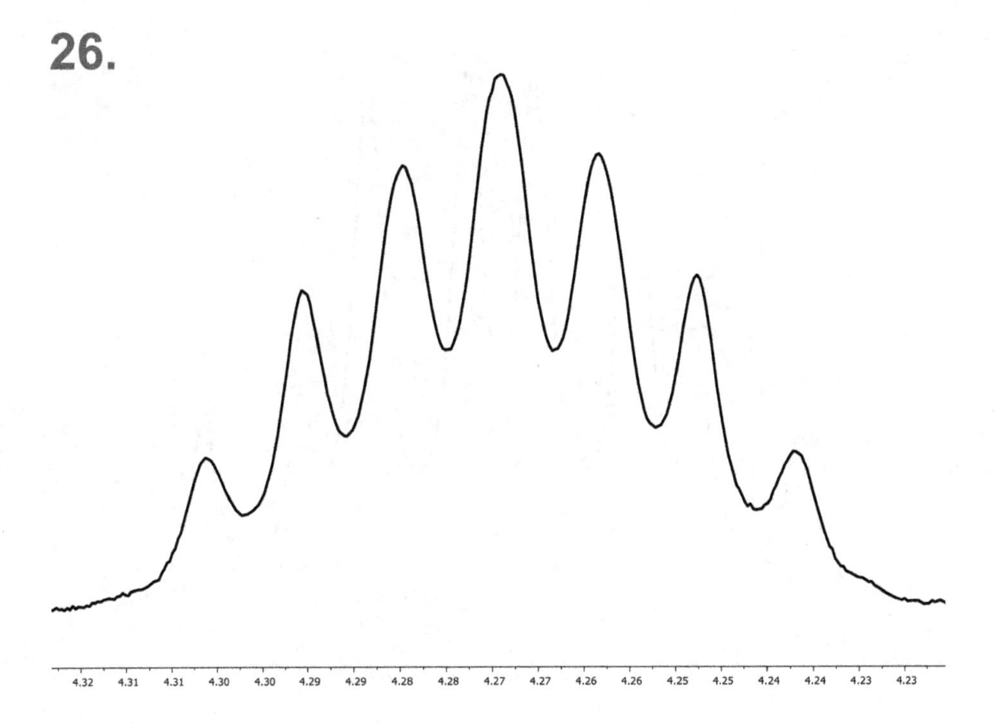

Figure 5.44: Advanced exercise 26.

27.

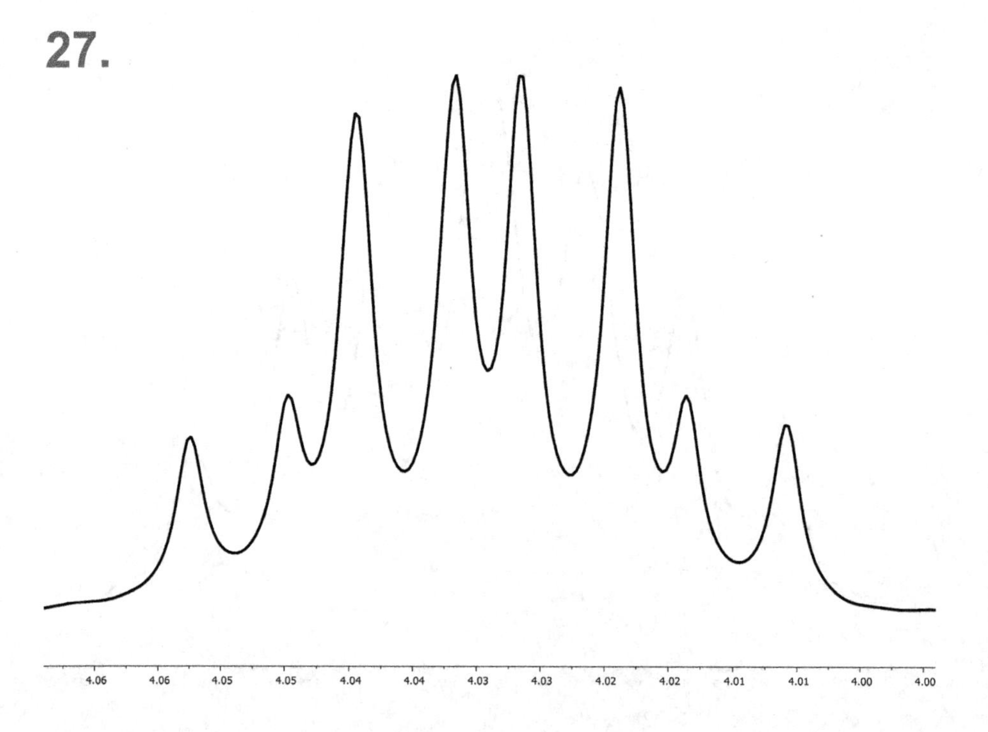

Figure 5.45: Advanced exercise 27.

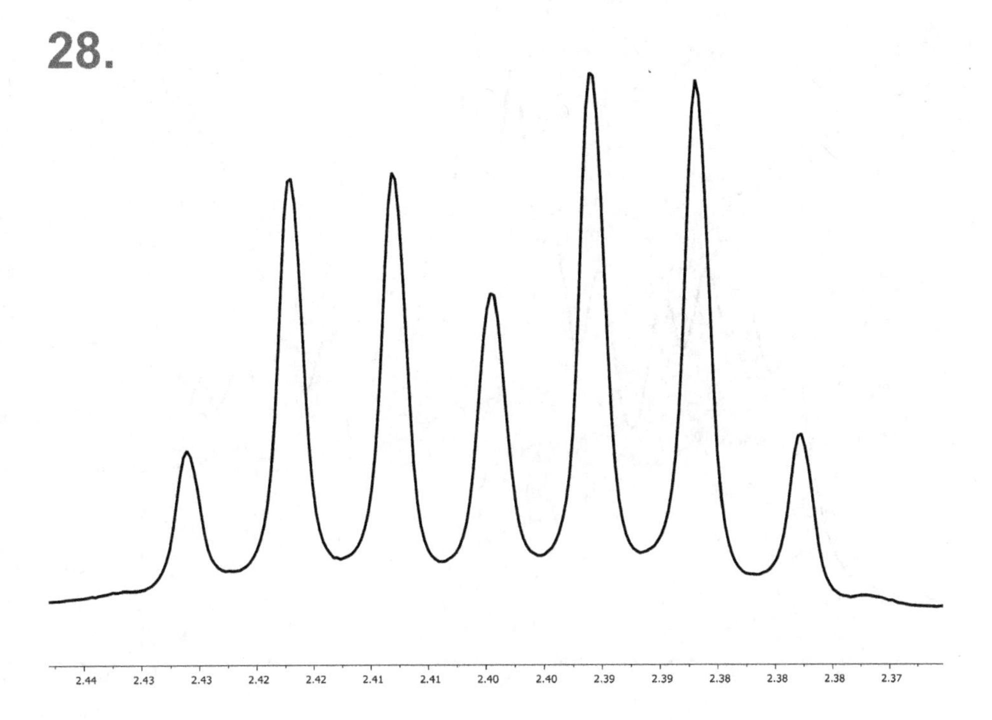

Figure 5.46: Advanced exercise 28.

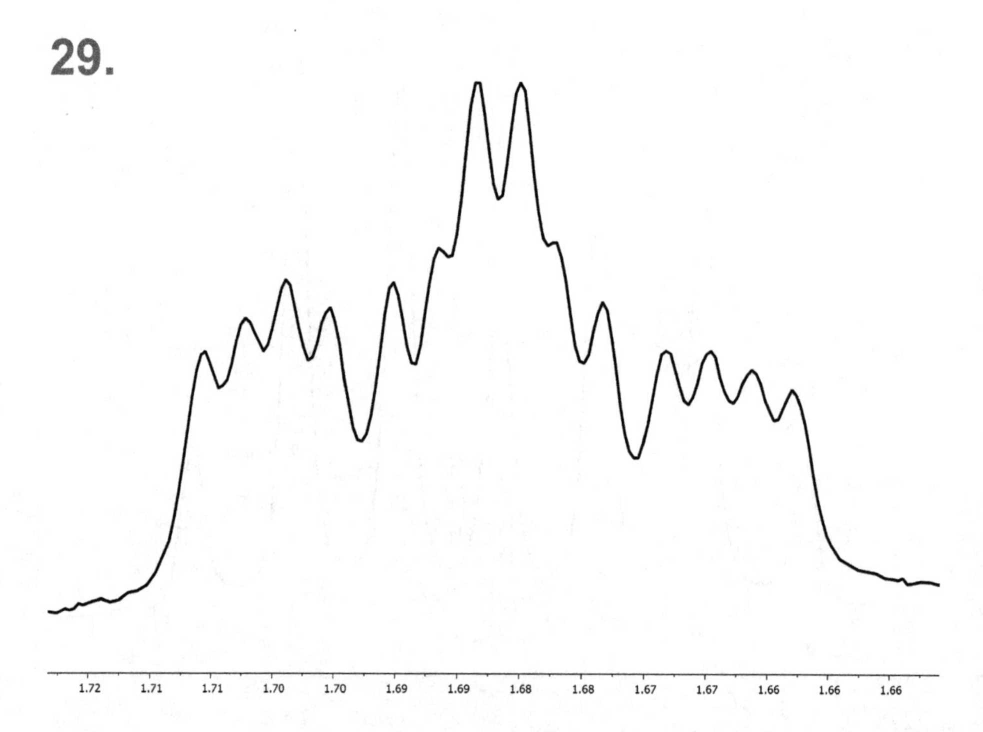

Figure 5.47: Advanced exercise 29.

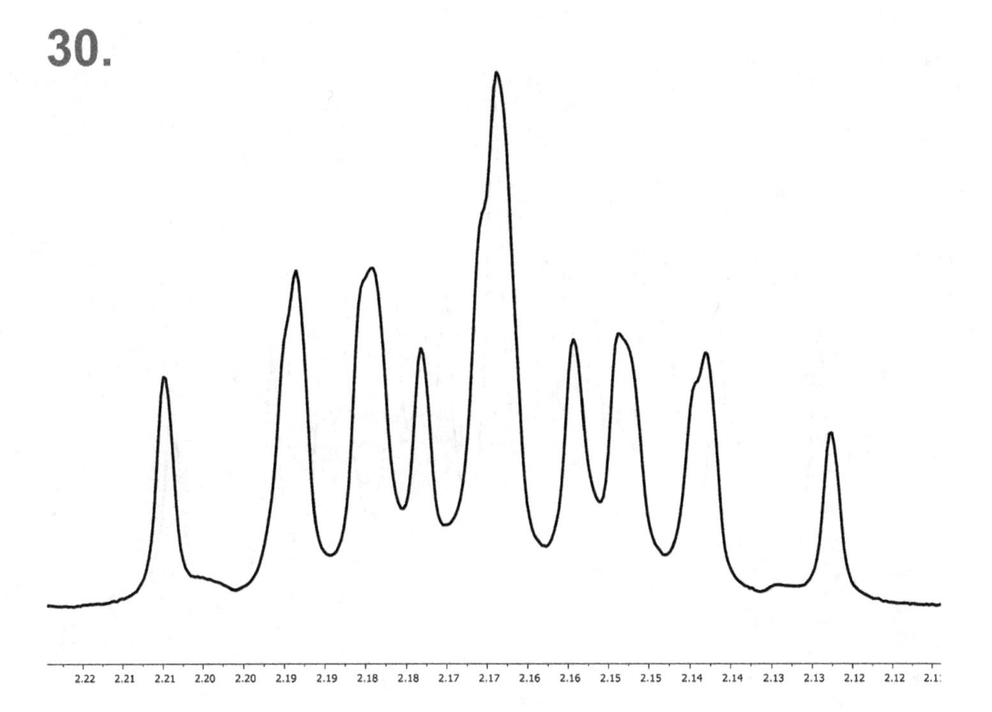

Figure 5.48: Advanced exercise 30.

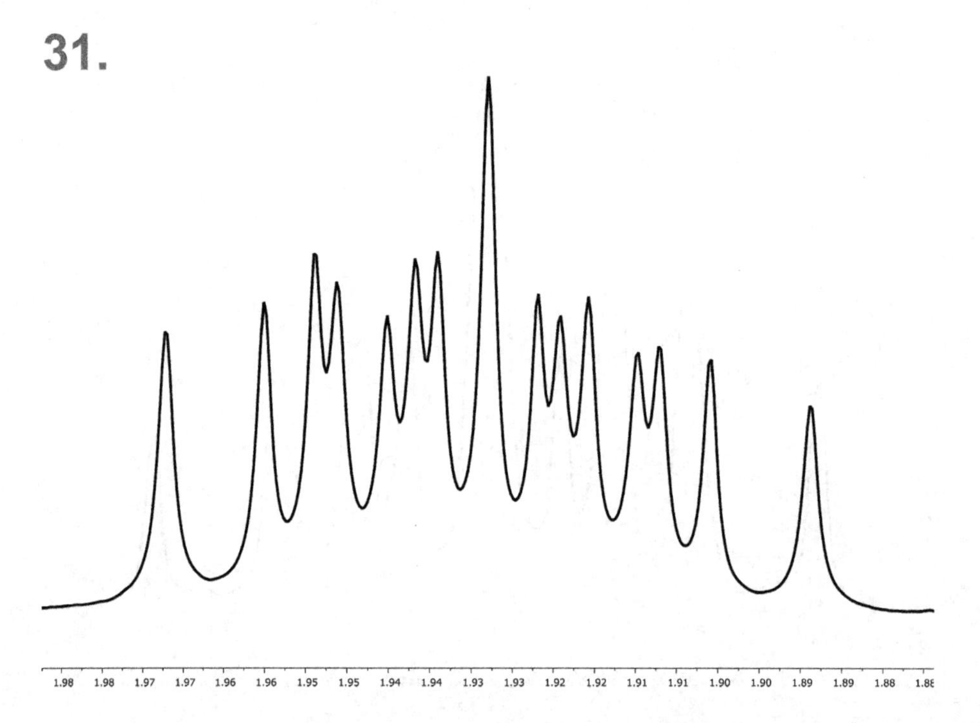

Figure 5.49: Advanced exercise 31.

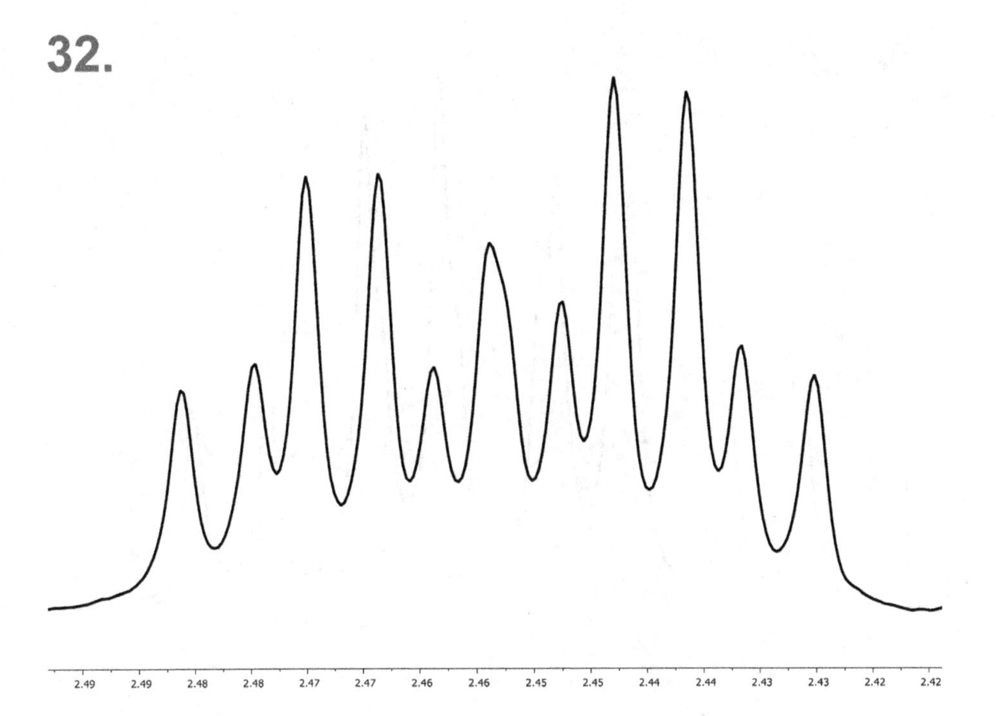

Figure 5.50: Advanced exercise 32.

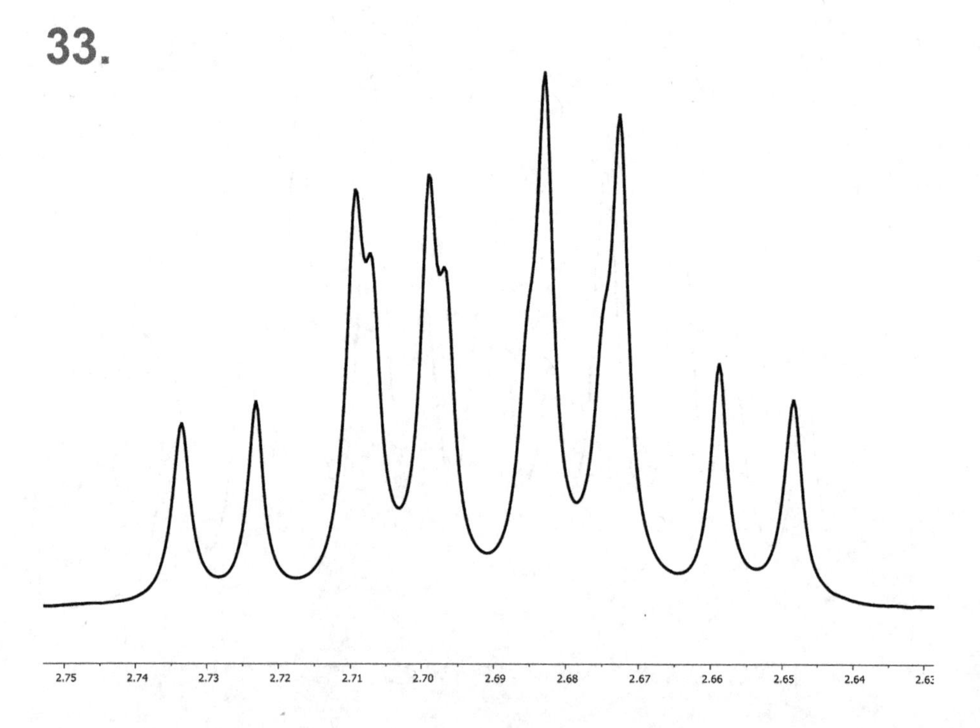

Figure 5.51: Advanced exercise 33.

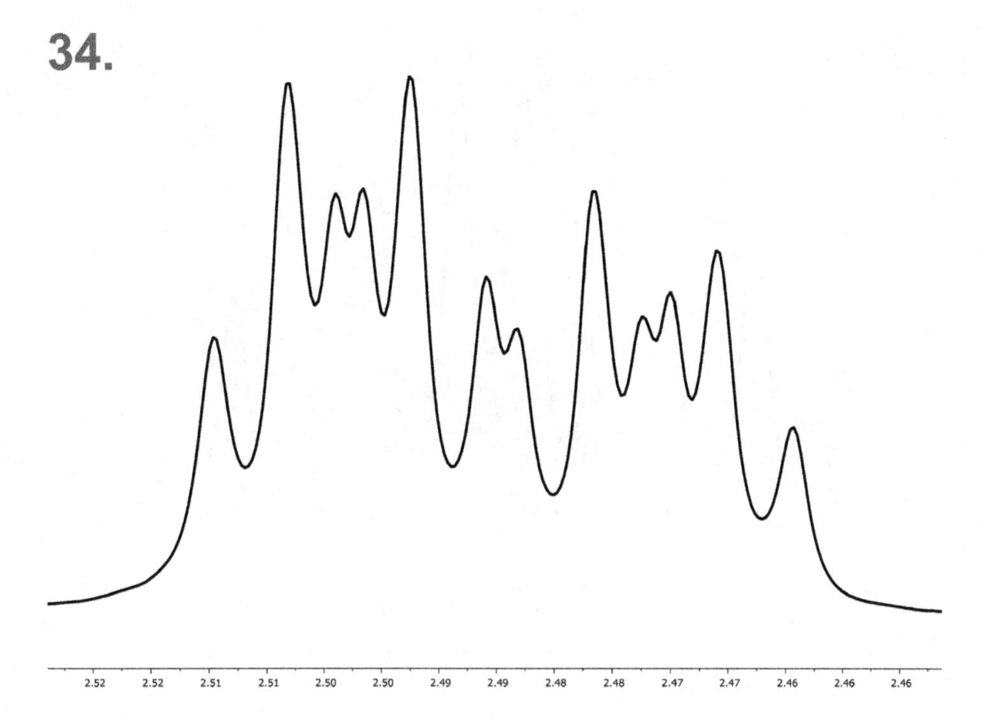

Figure 5.52: Advanced exercise 34.

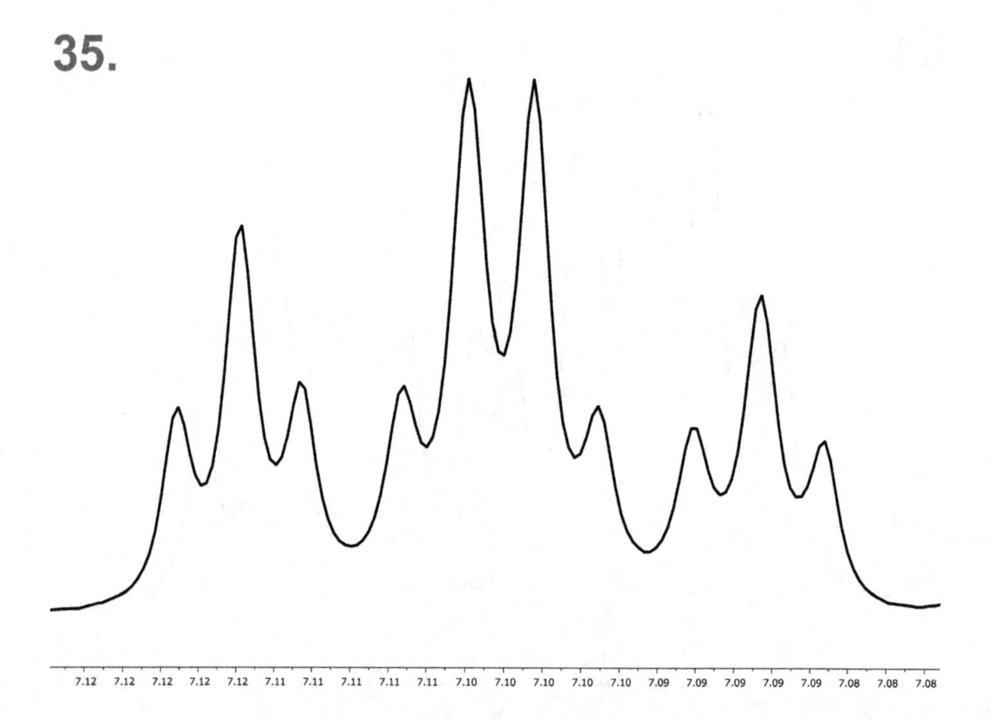

Figure 5.53: Advanced exercise 35.

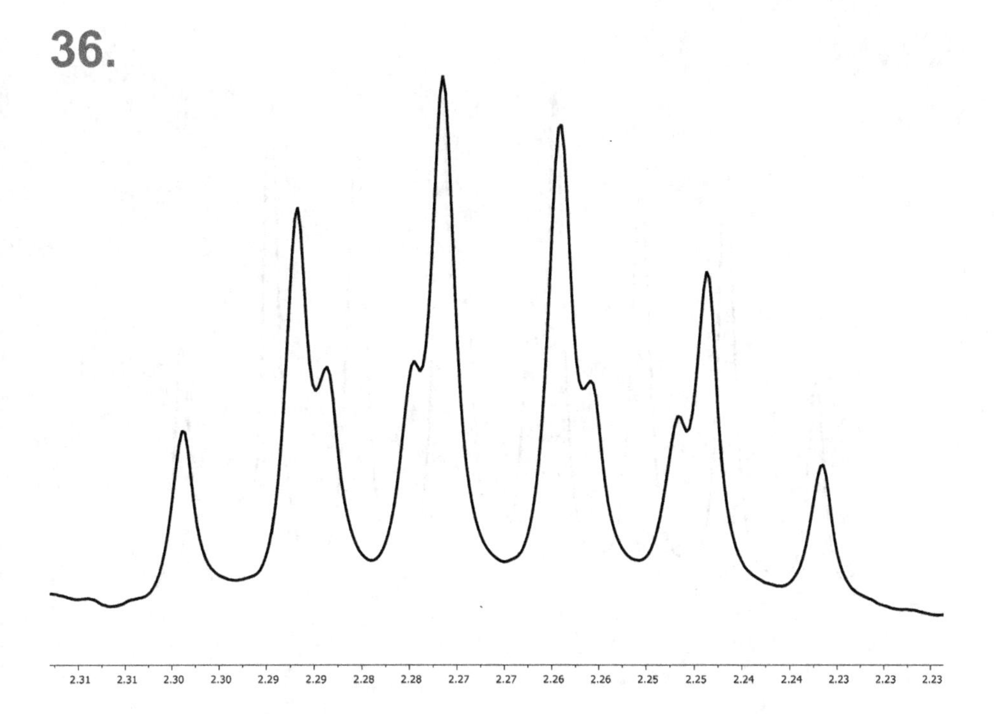

Figure 5.54: Advanced exercise 36.

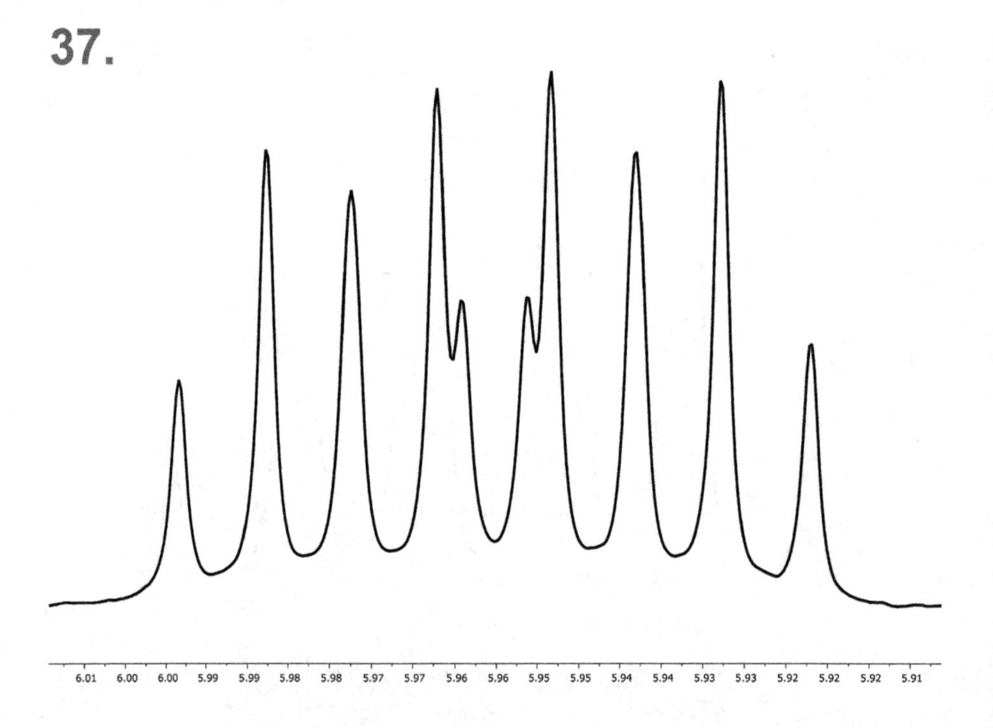

Figure 5.55: Advanced exercise 37.

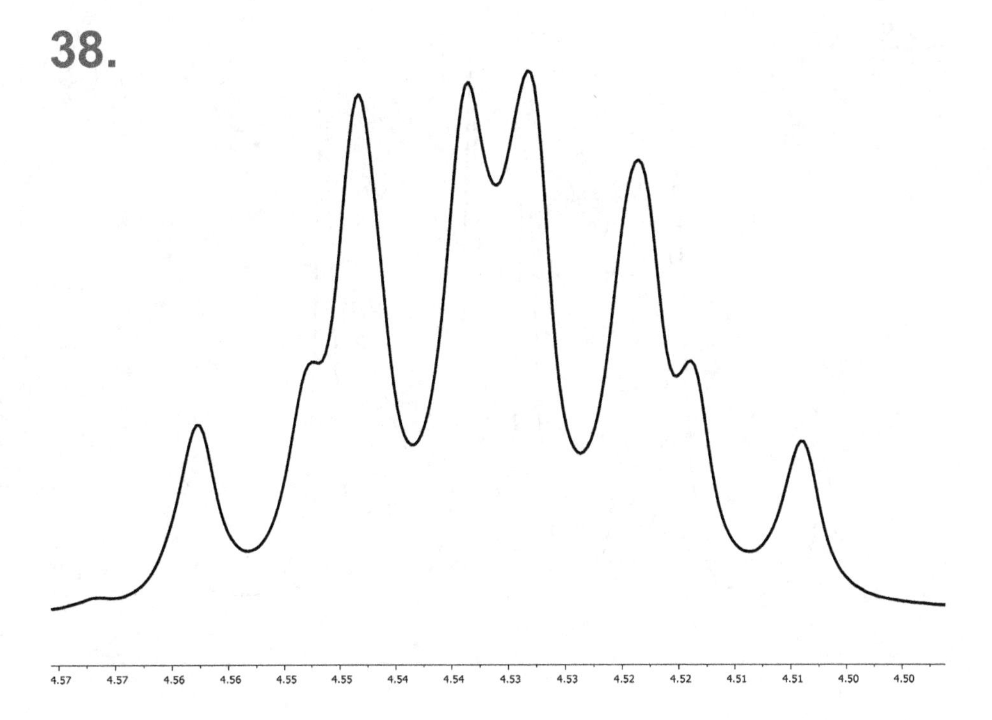

Figure 5.56: Advanced exercise 38.

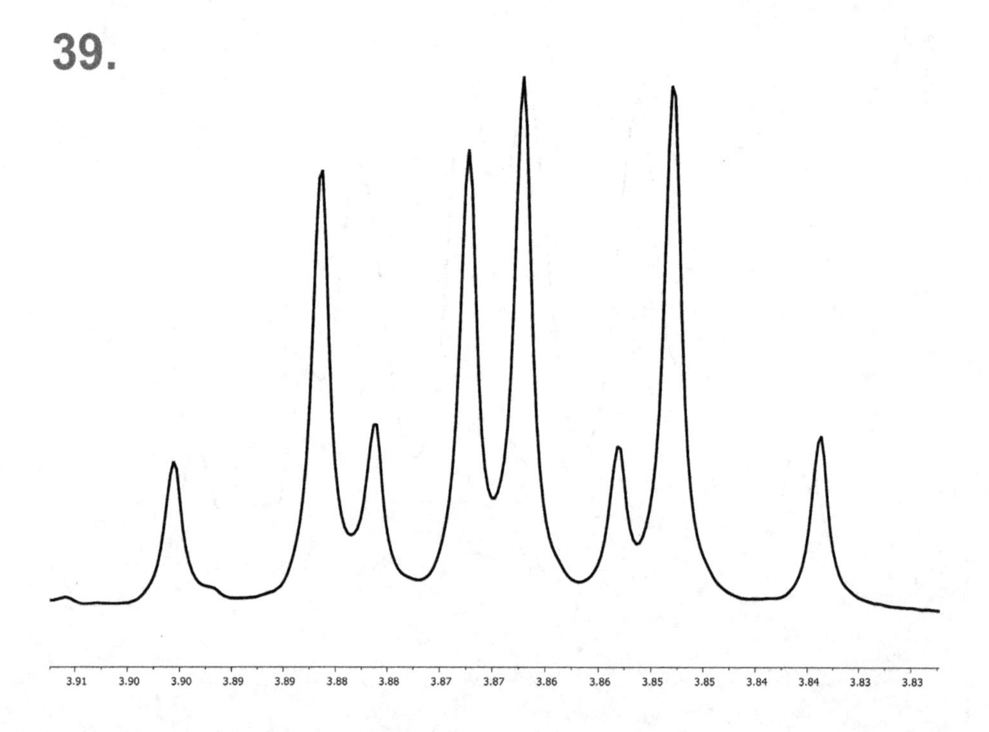

Figure 5.57: Advanced exercise 39.

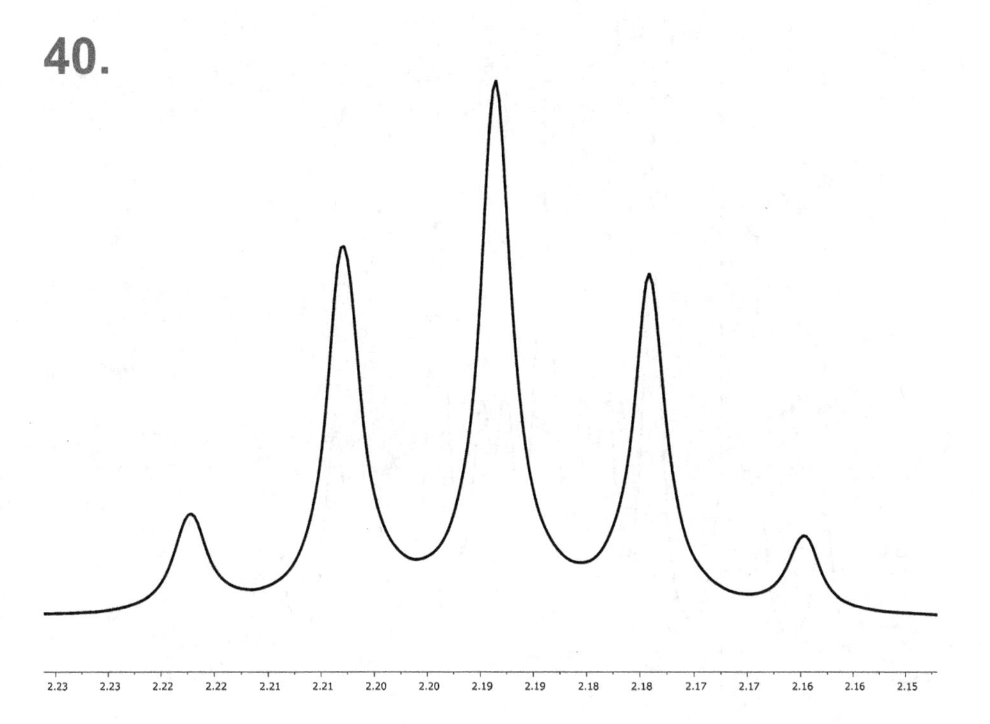

Figure 5.58: Advanced exercise 40.

5.3.2 Advanced Answers

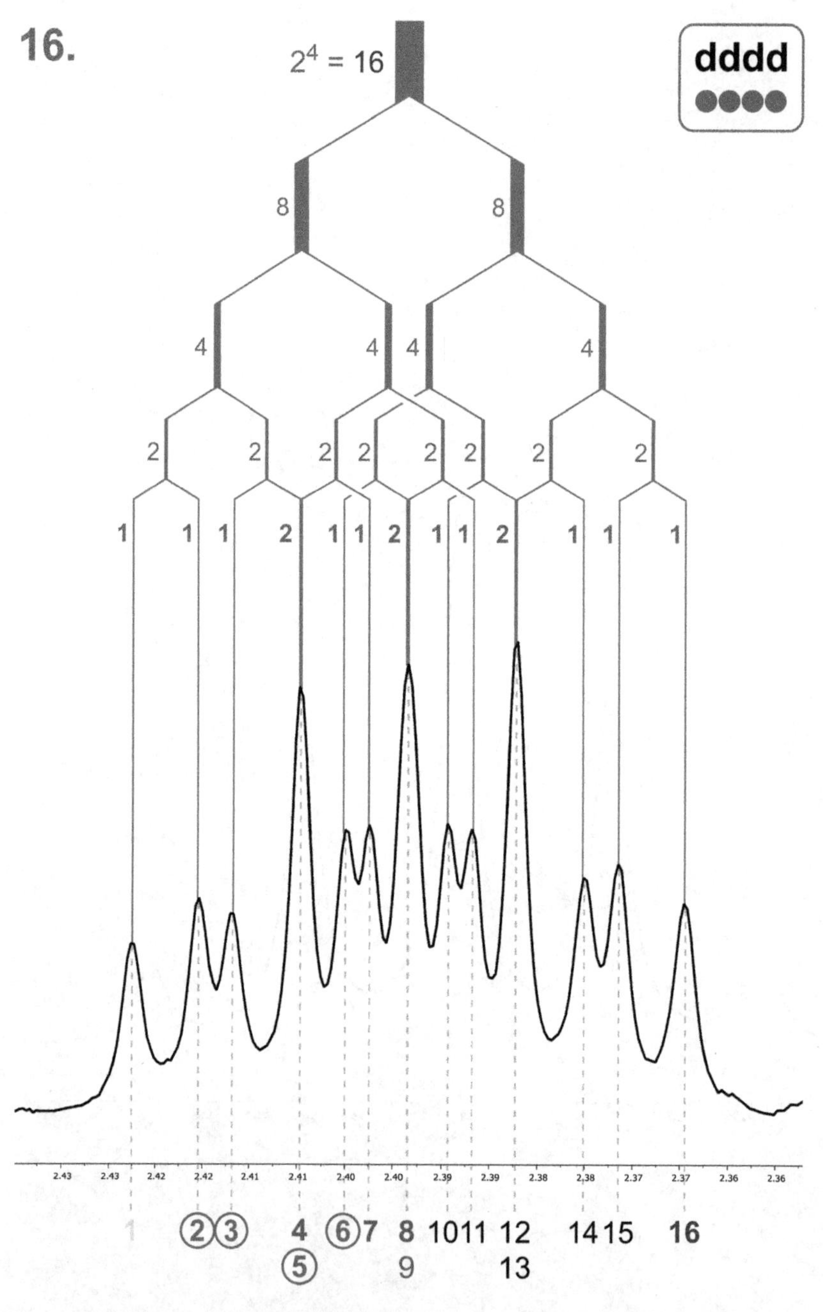

Figure 5.59: Advanced answer 16 [dddd].

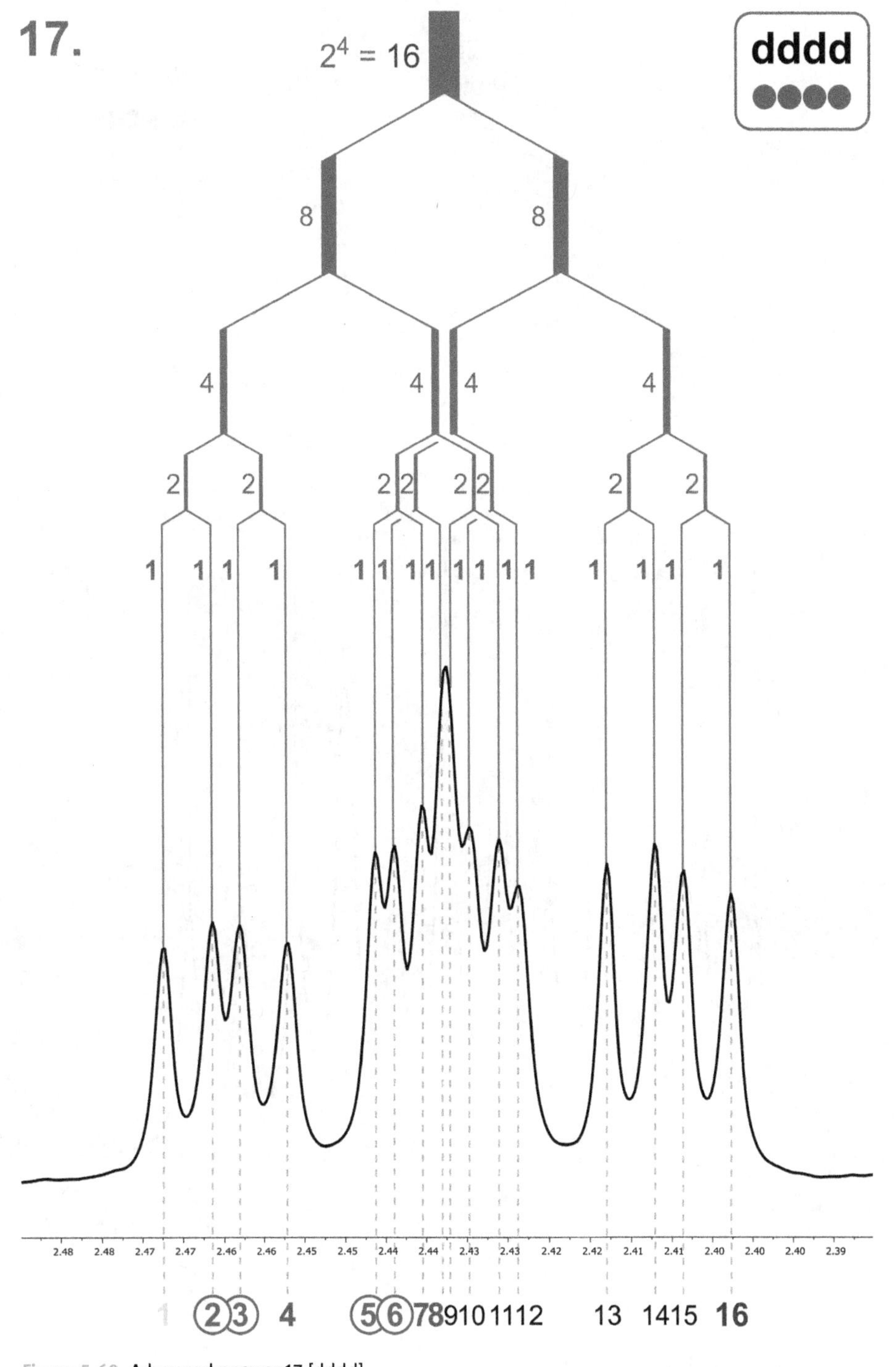

Figure 5.60: Advanced answer 17 [dddd].

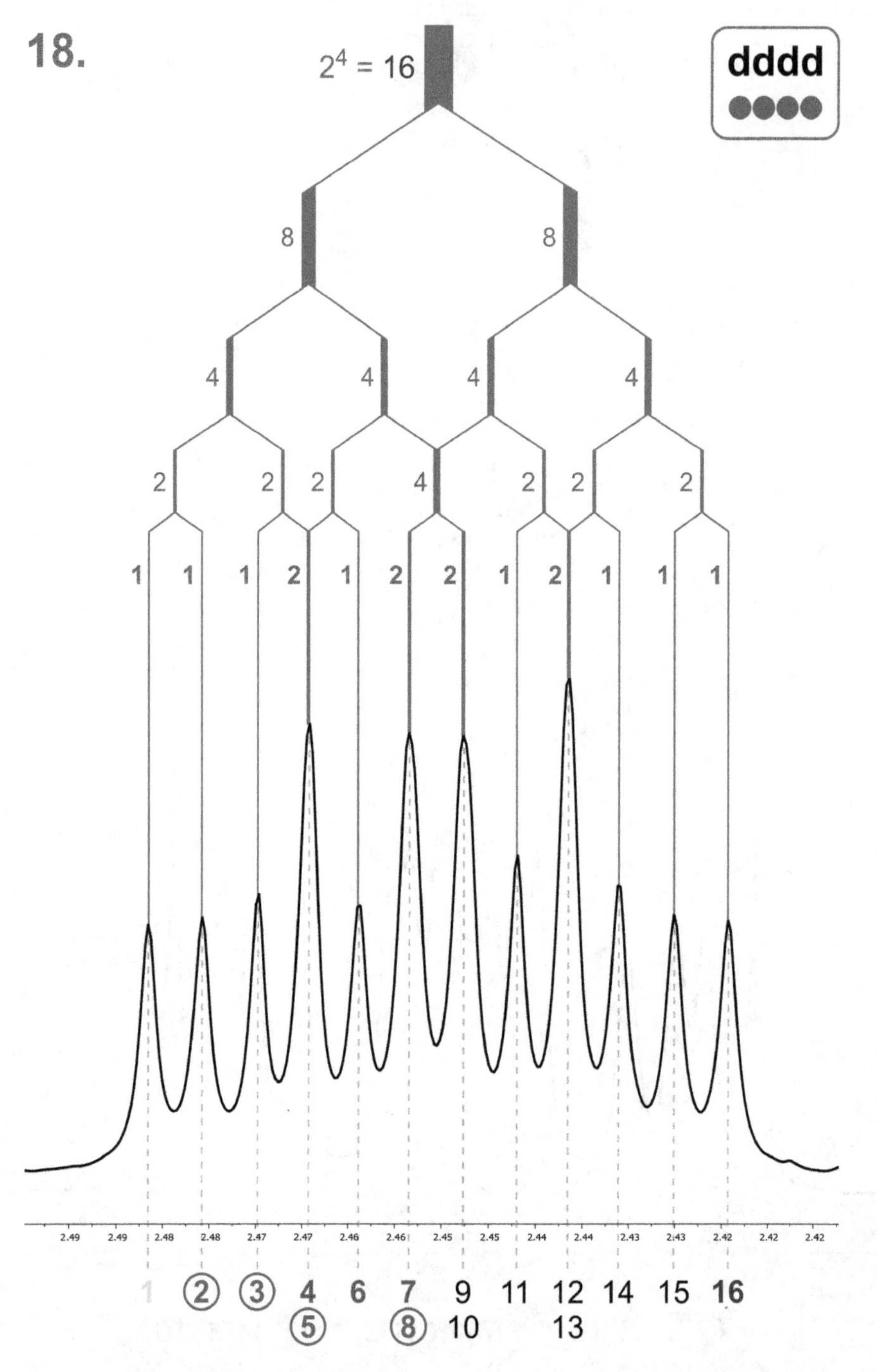

Figure 5.61: Advanced answer 18 [dddd].

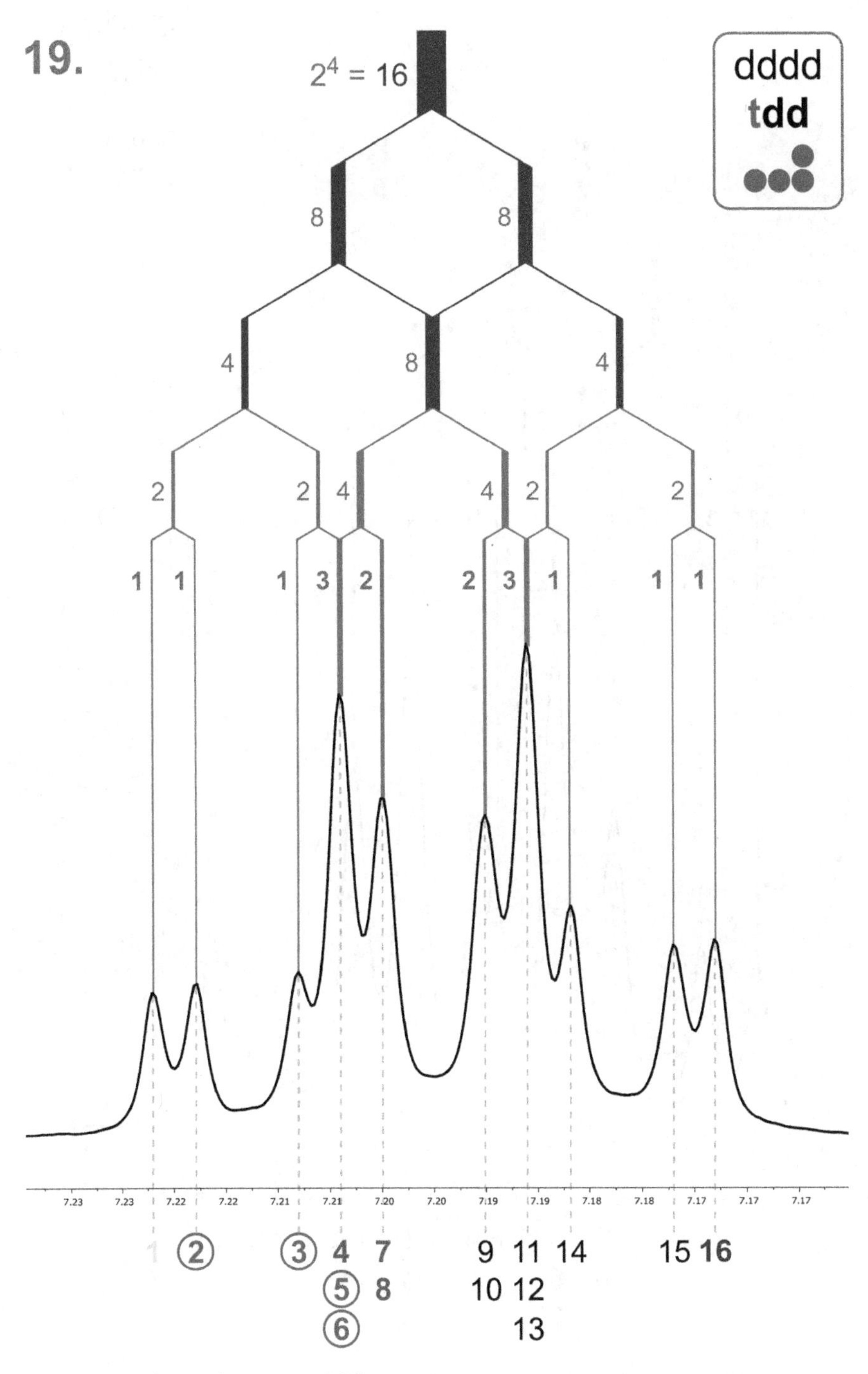

Figure 5.62: Advanced answer 19 [tdd].

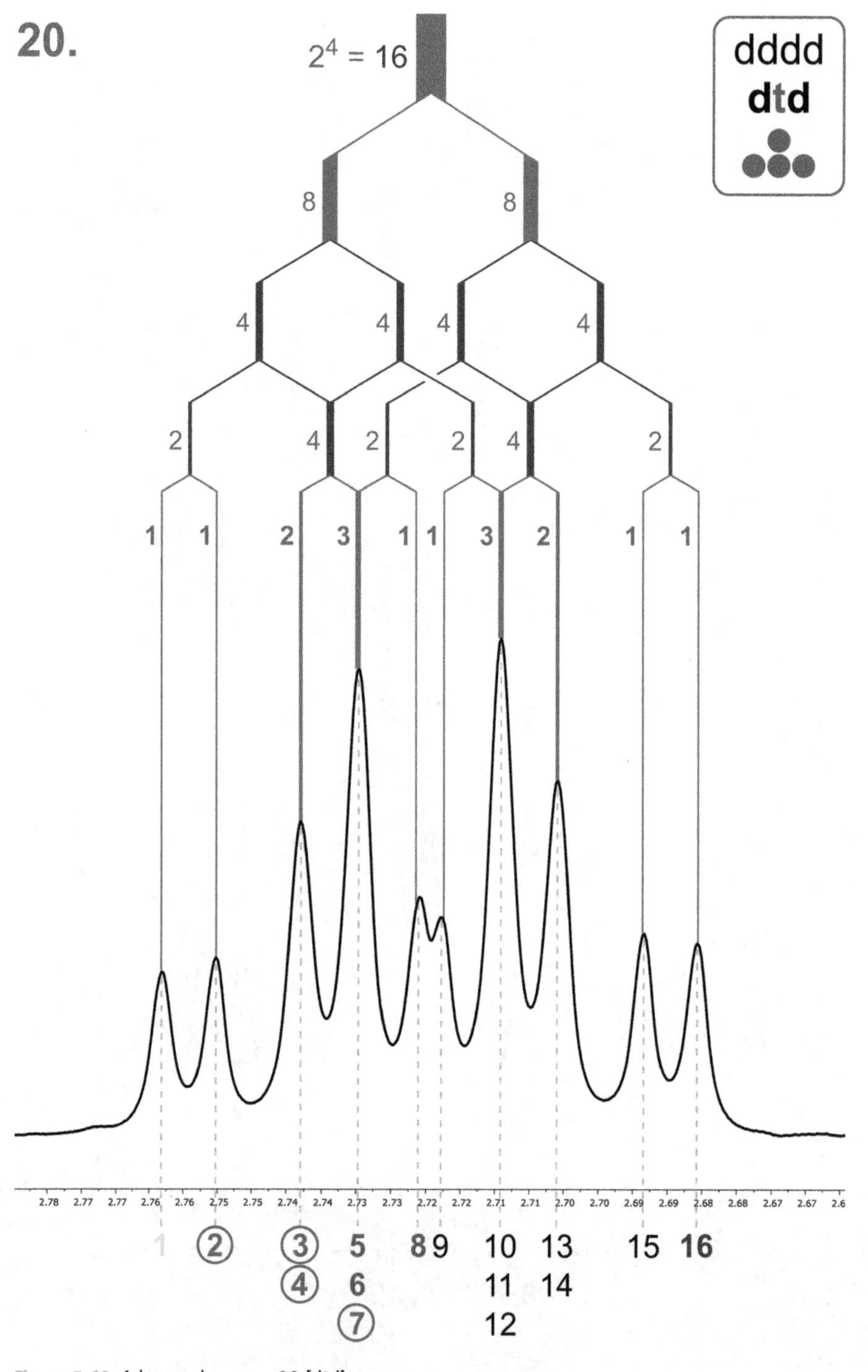

Figure 5.63: Advanced answer 20 [dtd].

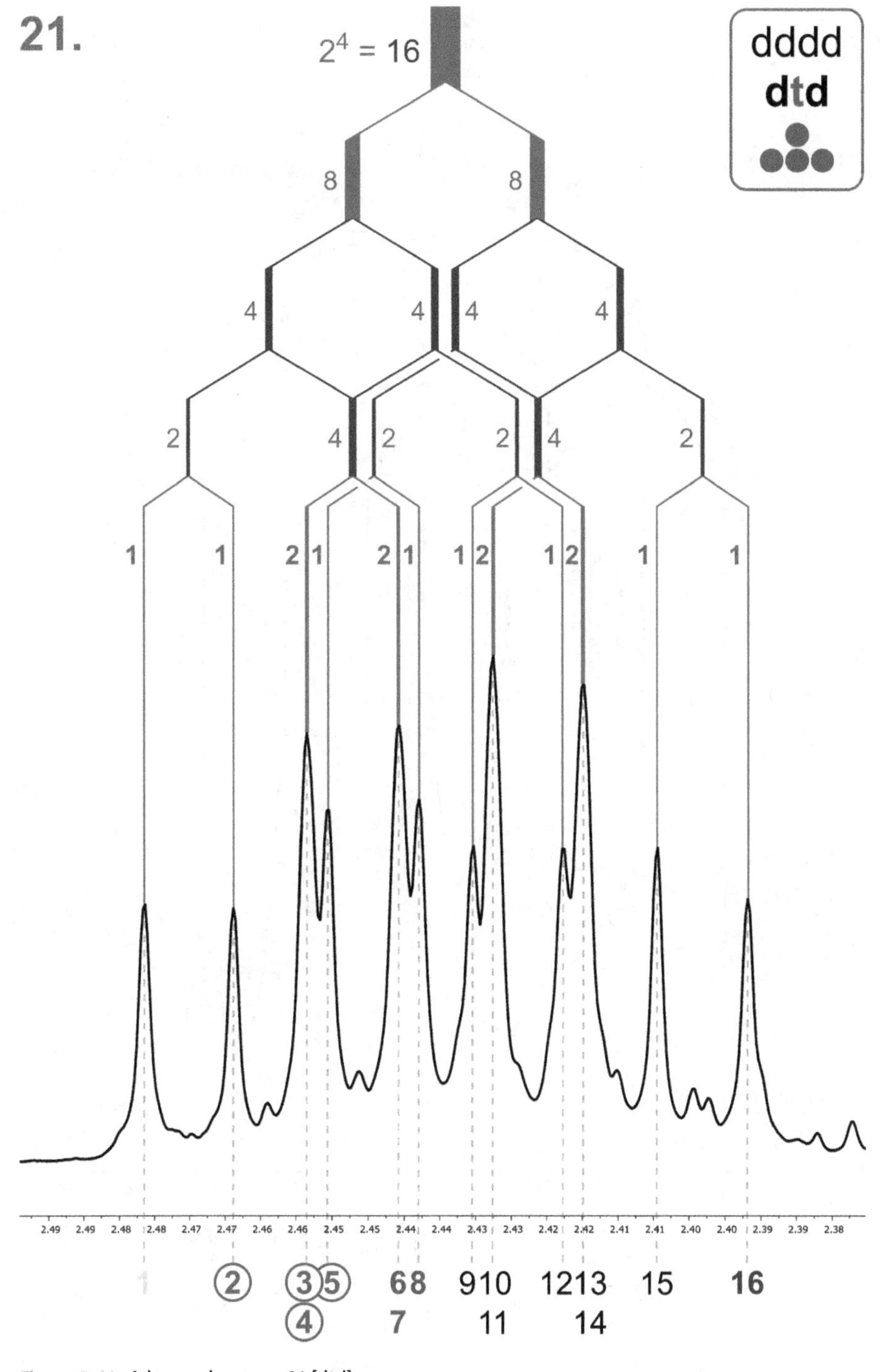

Figure 5.64: Advanced answer 21 [dtd].

22.

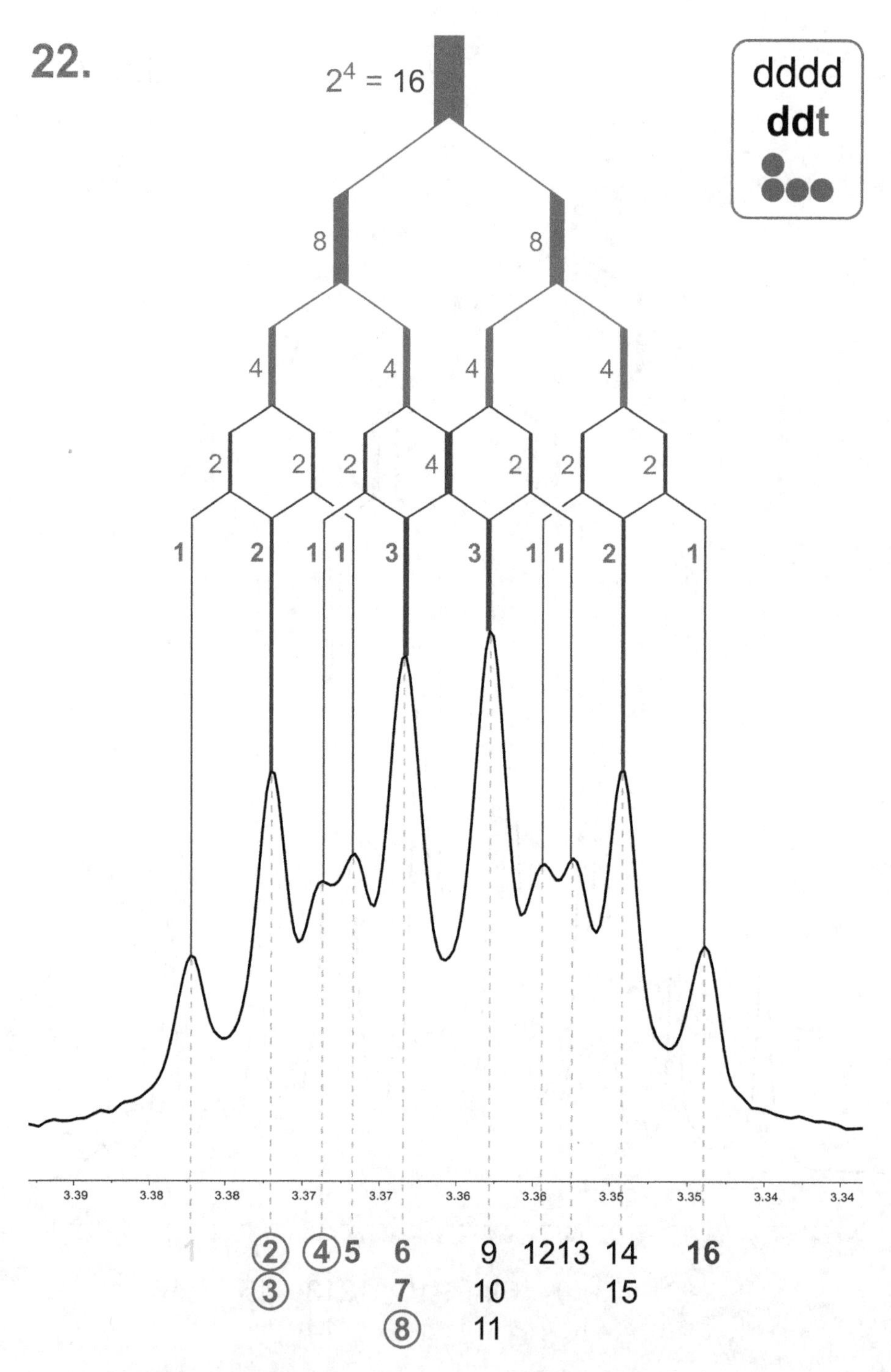

Figure 5.65: Advanced answer 22 [ddt].

23.

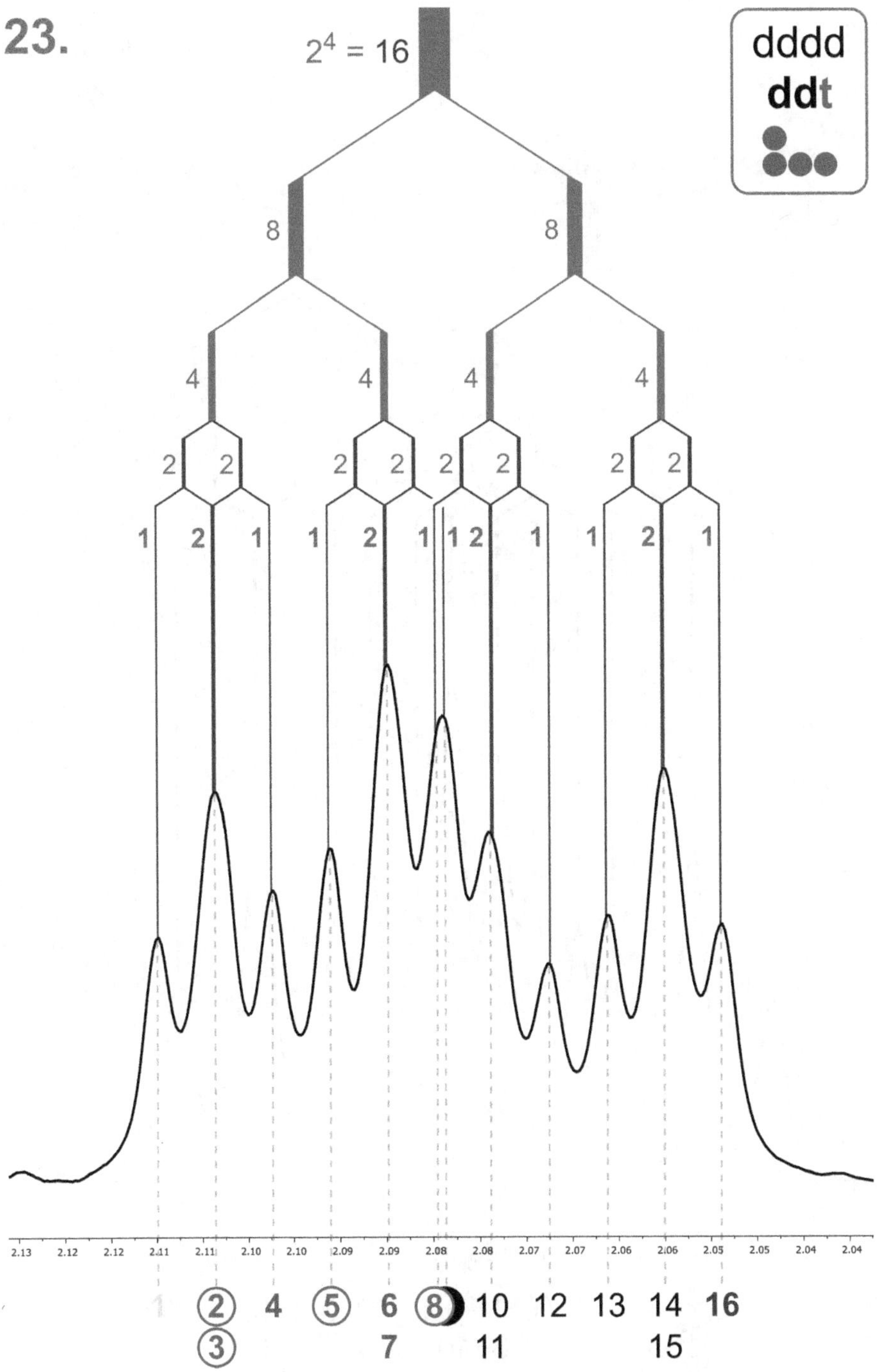

Figure 5.66: Advanced answer 23 [ddt].

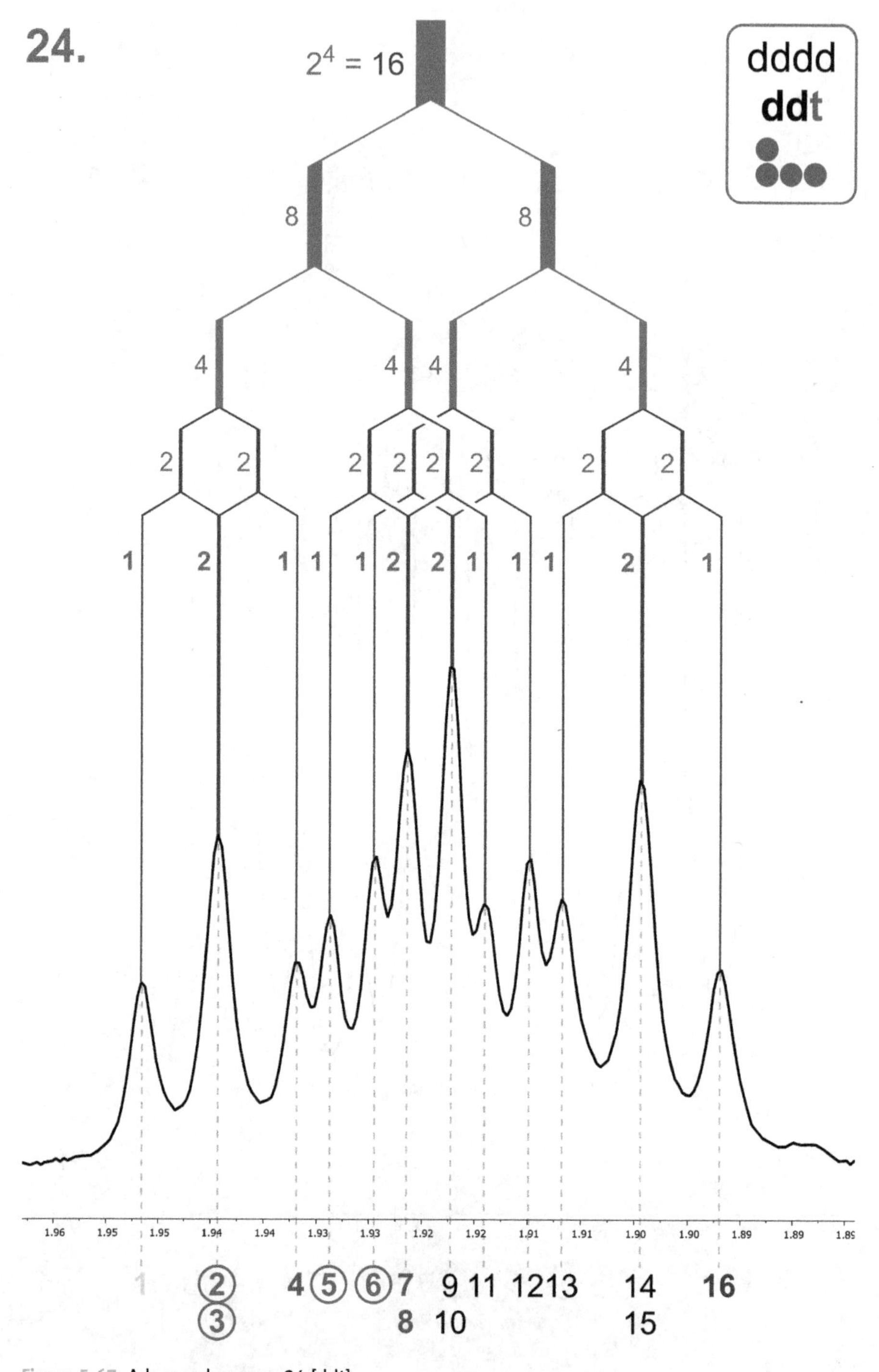

Figure 5.67: Advanced answer 24 [ddt].

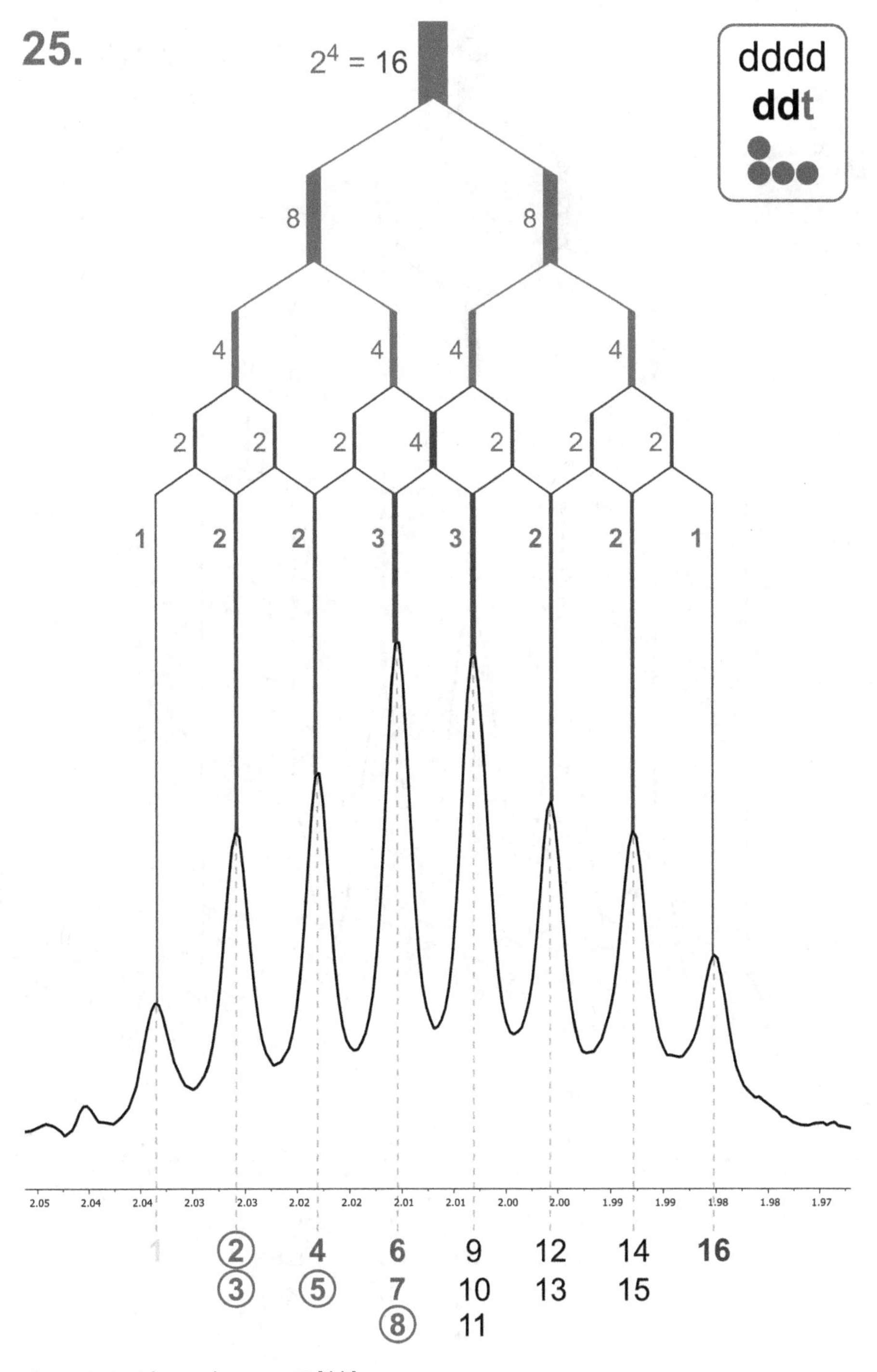

Figure 5.68: Advanced answer 25 [ddt].

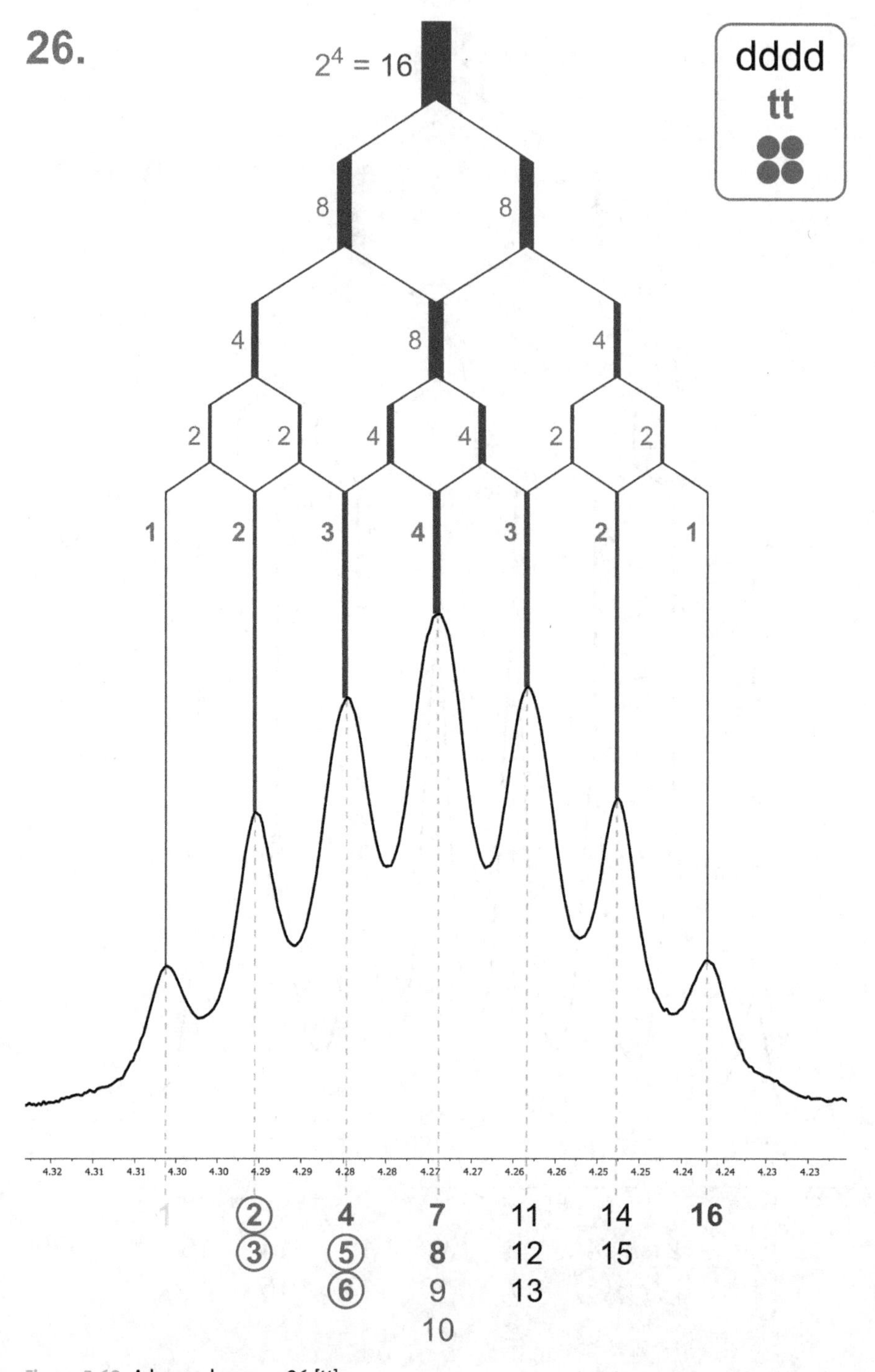

Figure 5.69: Advanced answer 26 [tt].

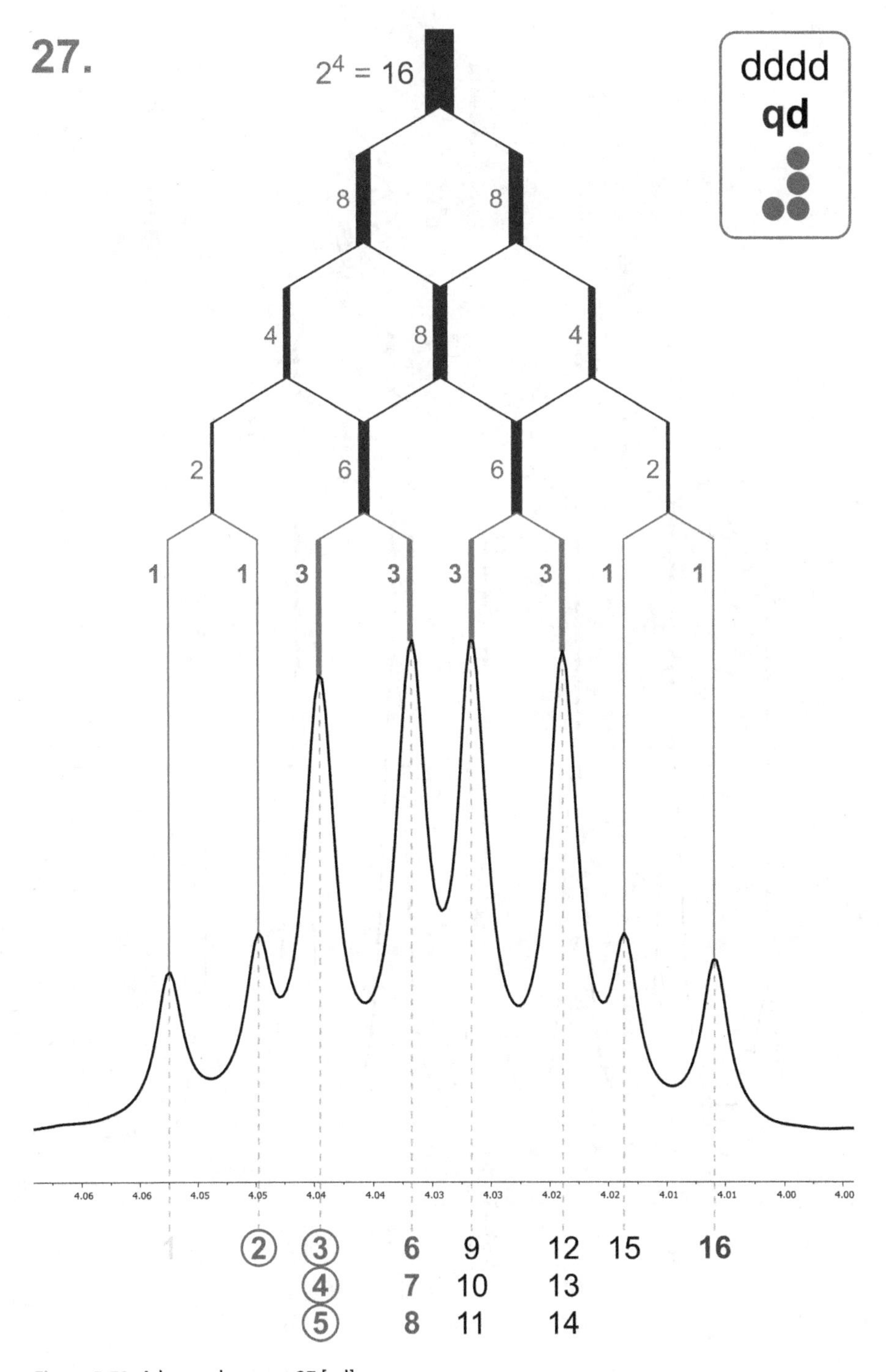

Figure 5.70: Advanced answer 27 [qd].

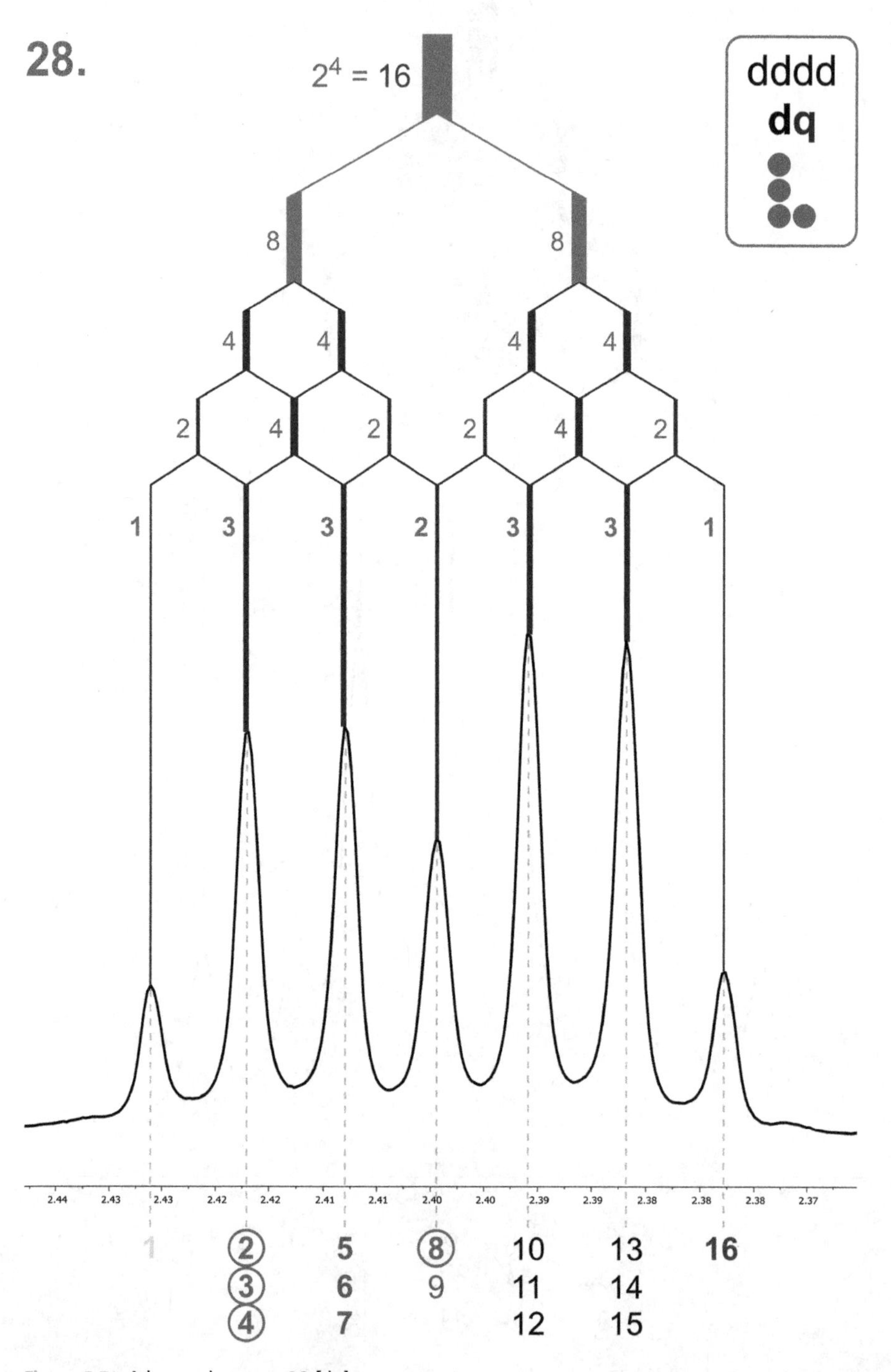

Figure 5.71: Advanced answer 28 [dq].

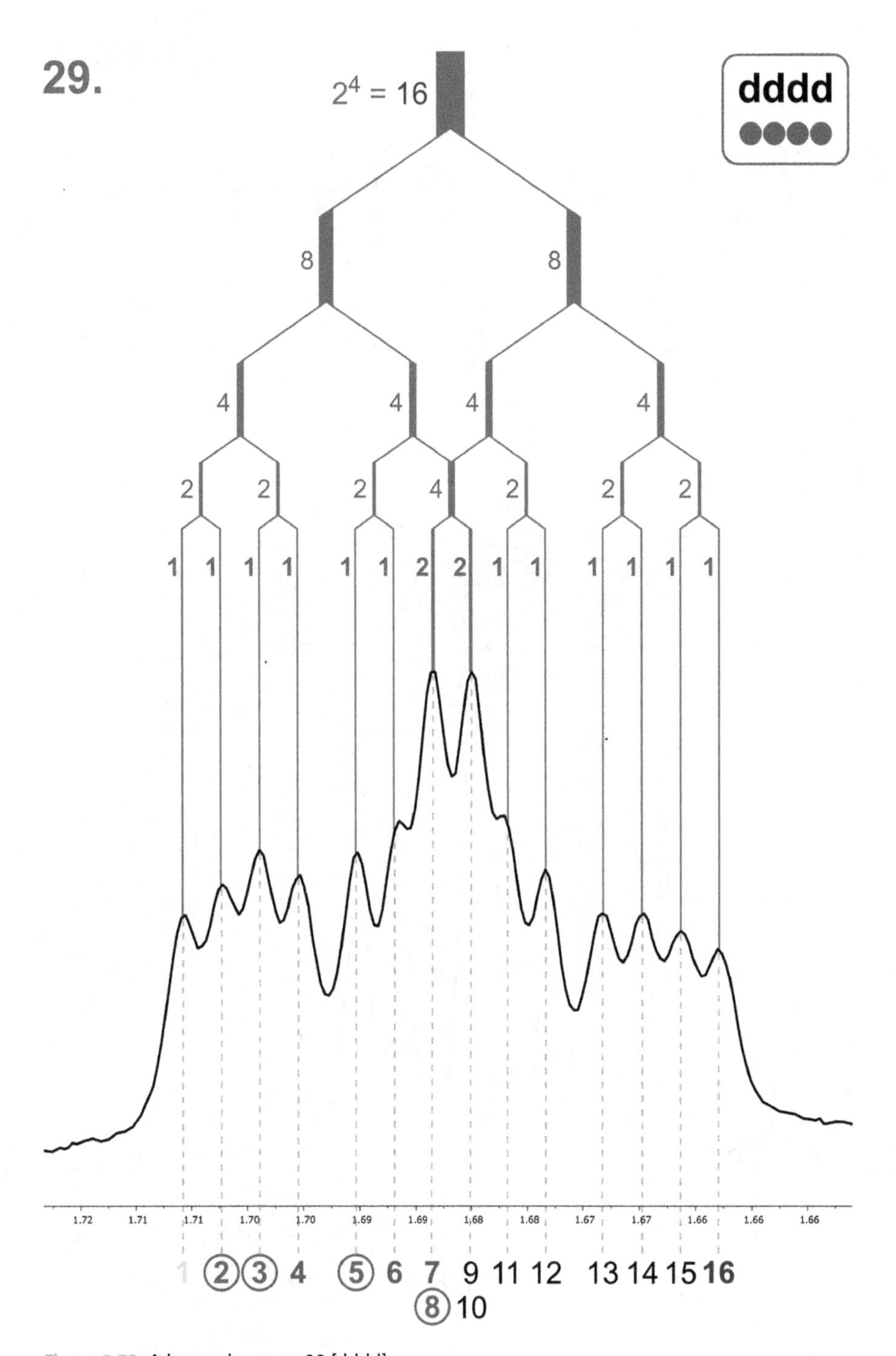

Figure 5.72: Advanced answer 29 [dddd].

30.

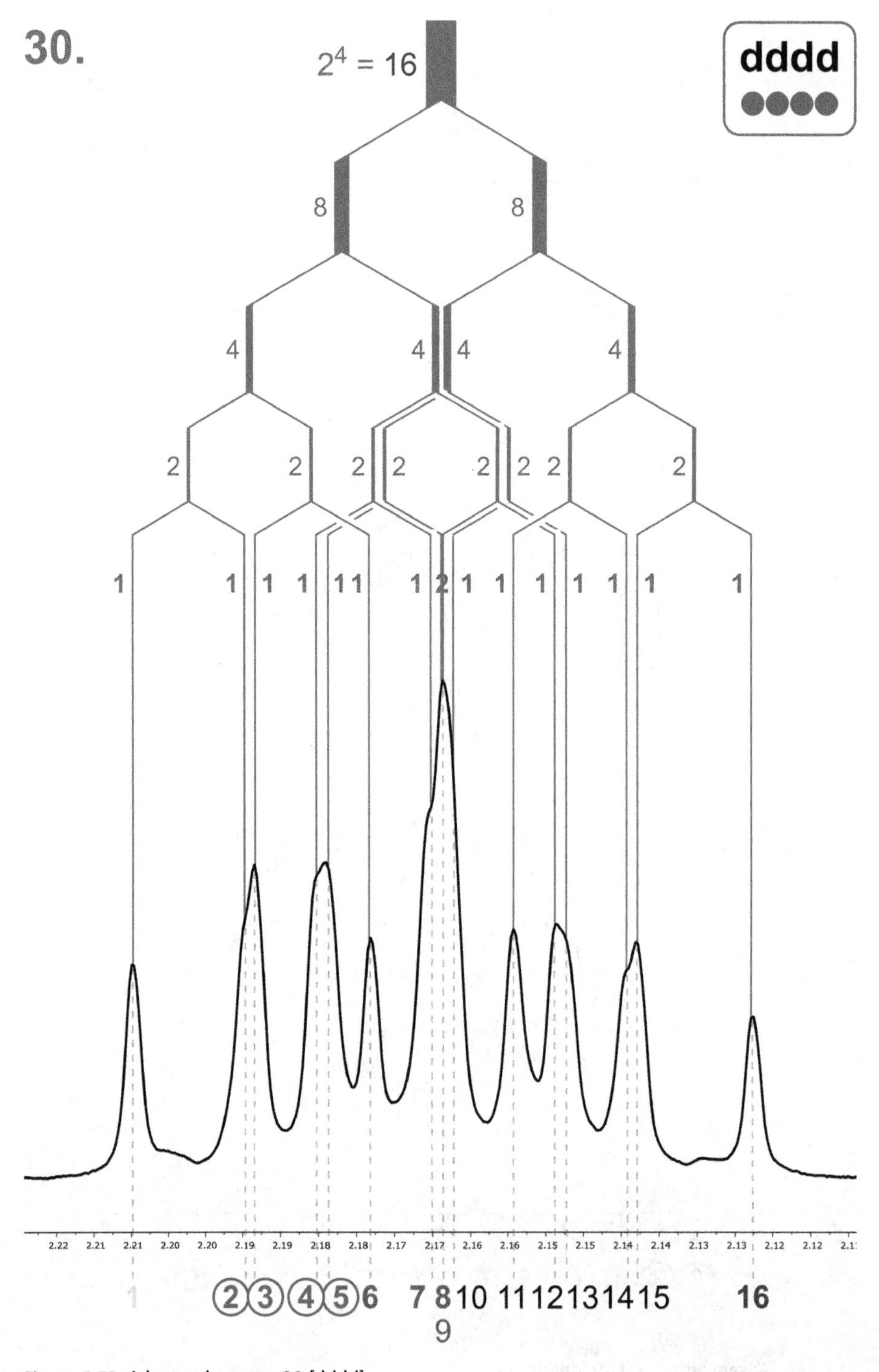

Figure 5.73: Advanced answer 30 [dddd].

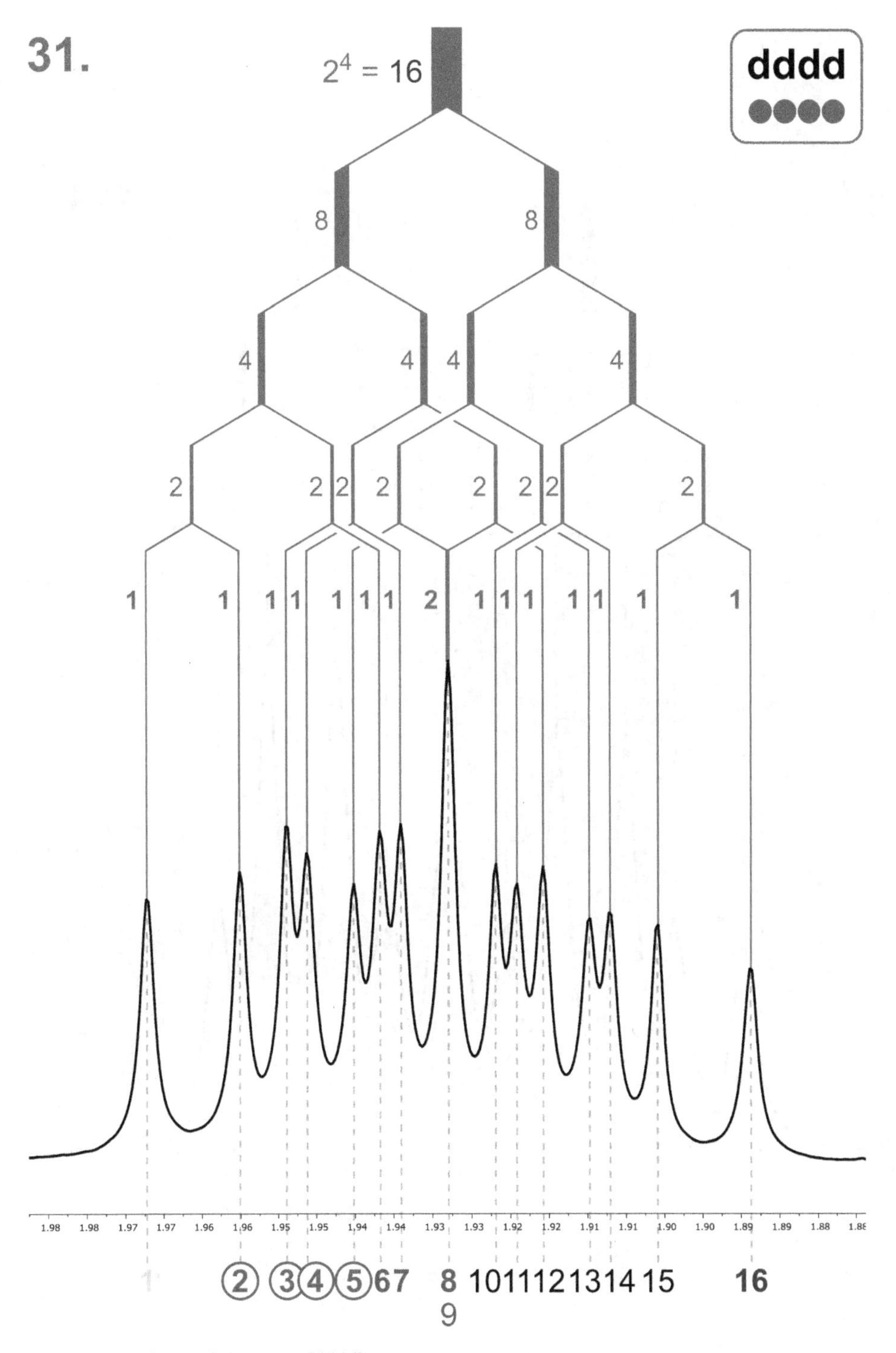

Figure 5.74: Advanced answer 31 [dddd].

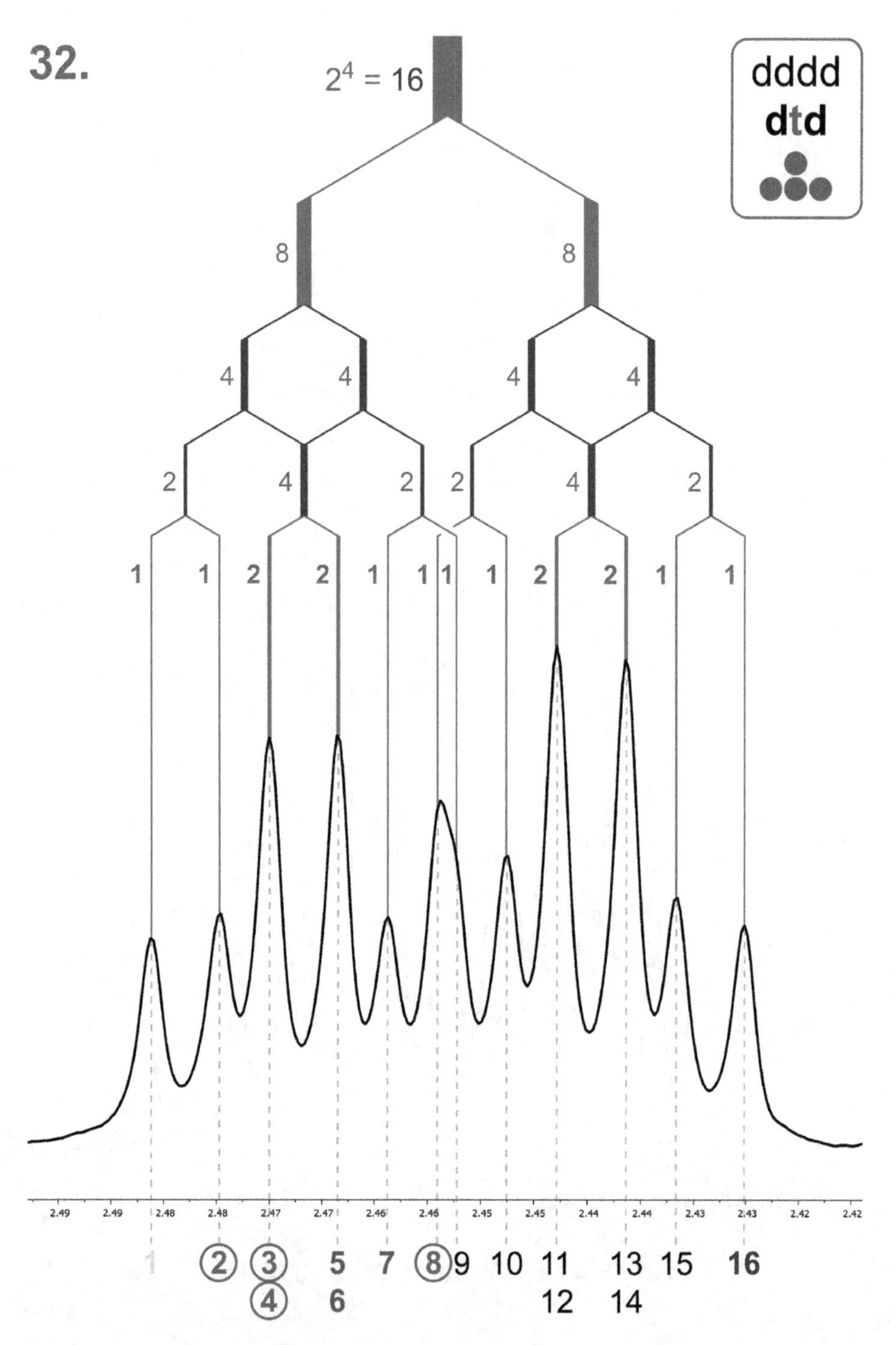

Figure 5.75: Advanced answer 32 [dtd].

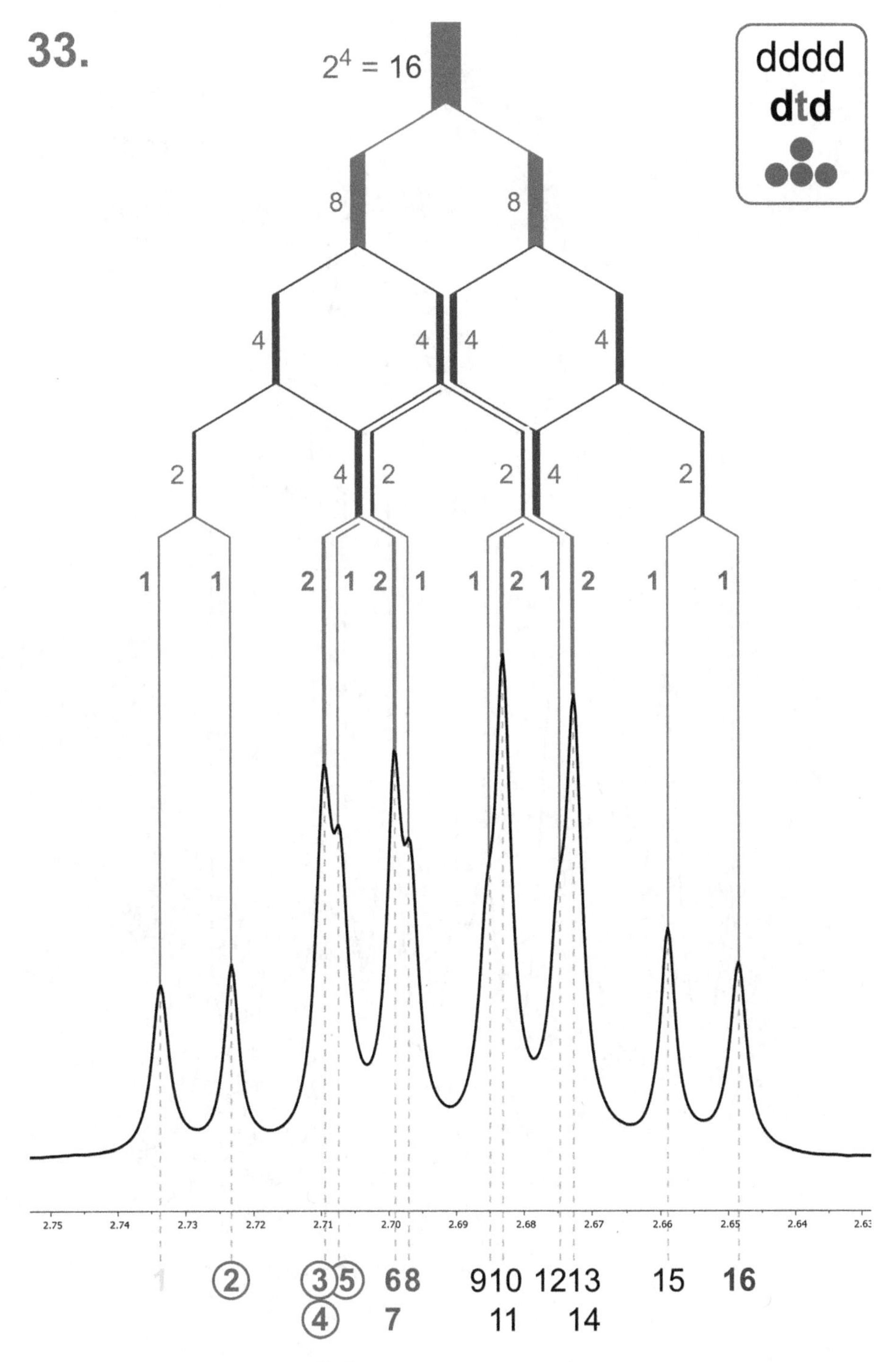

Figure 5.76: Advanced answer 33 [dtd].

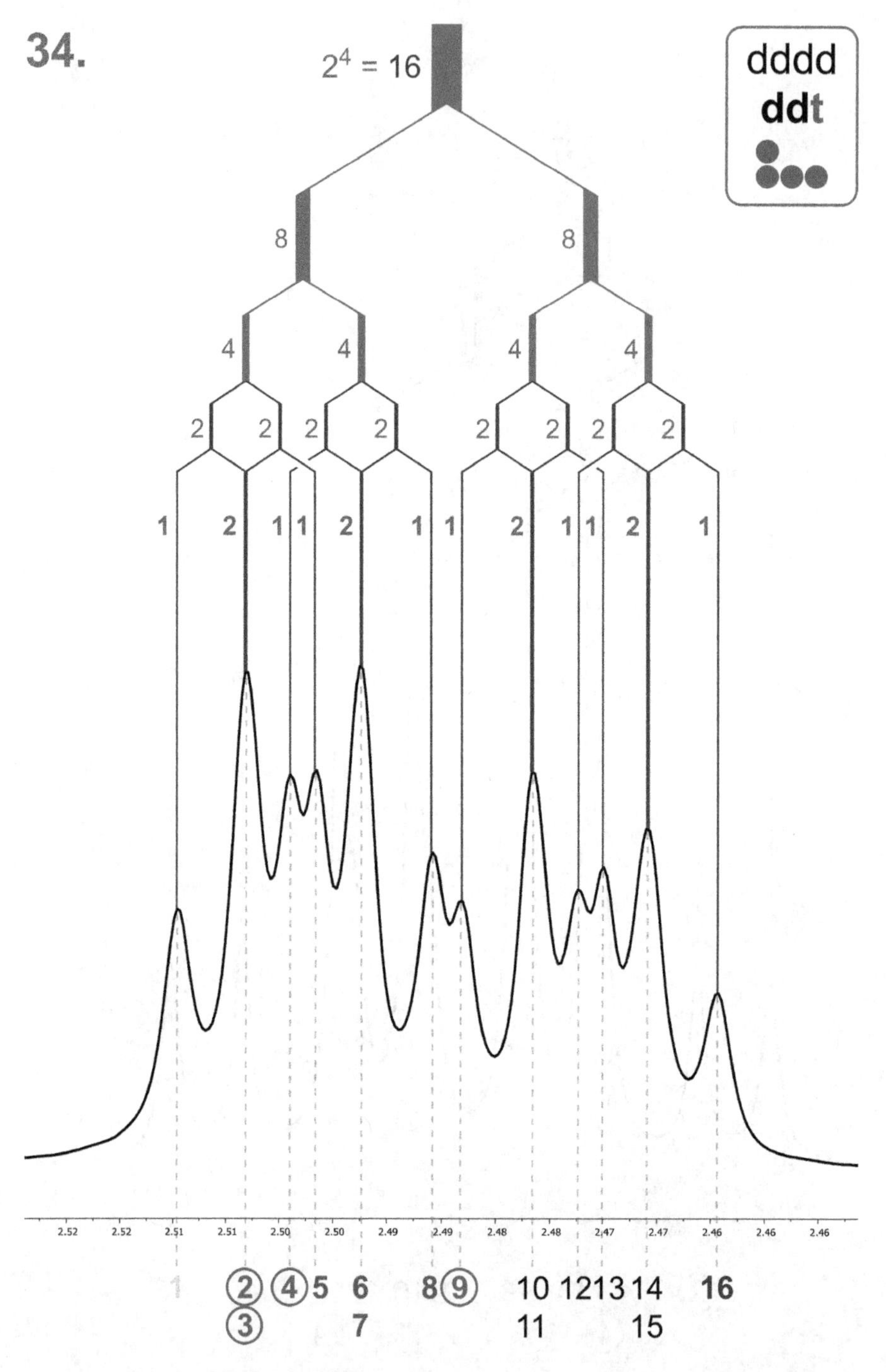

Figure 5.77: Advanced answer 34 [ddt].

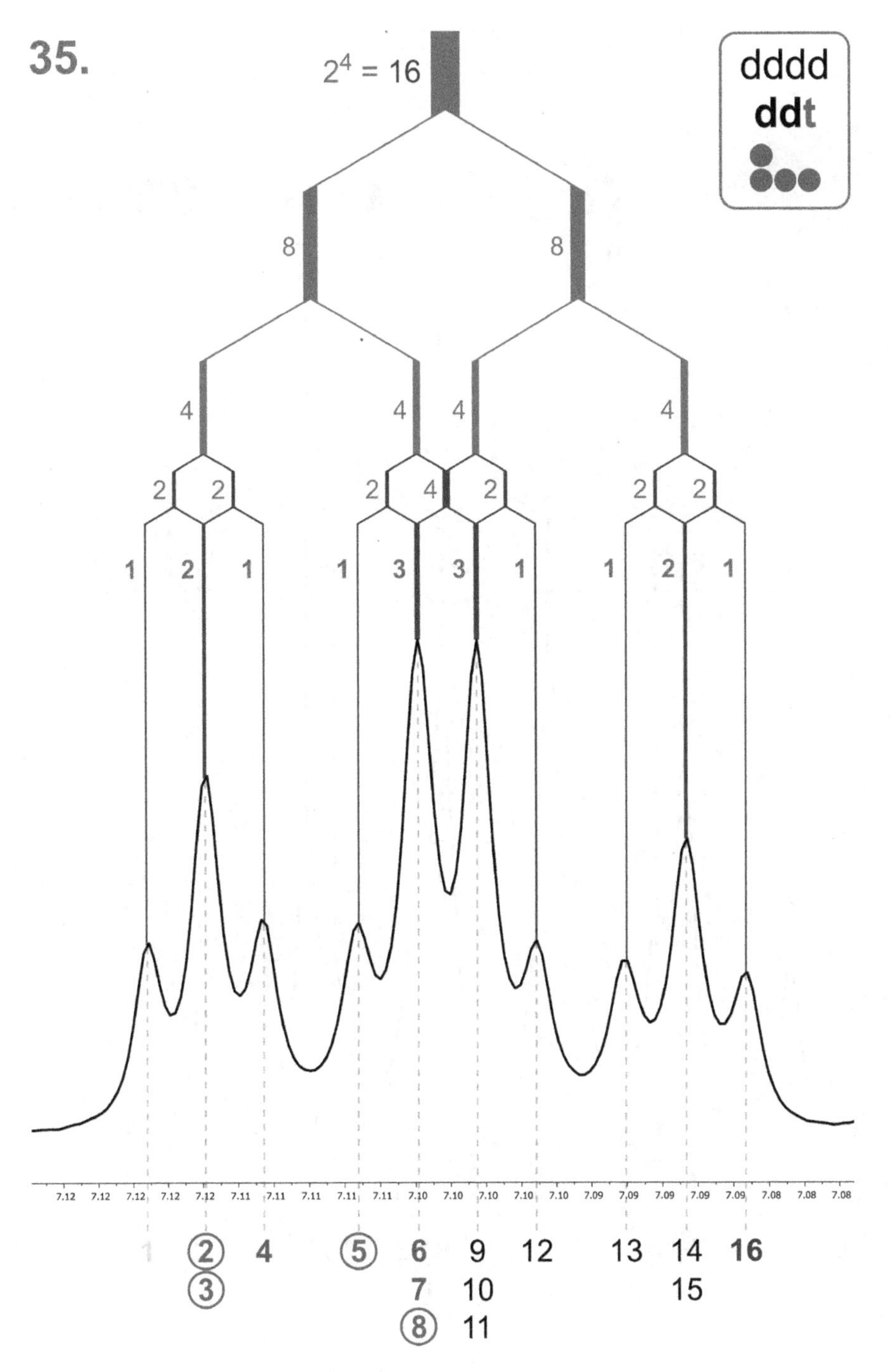

Figure 5.78: Advanced answer 35 [ddt].

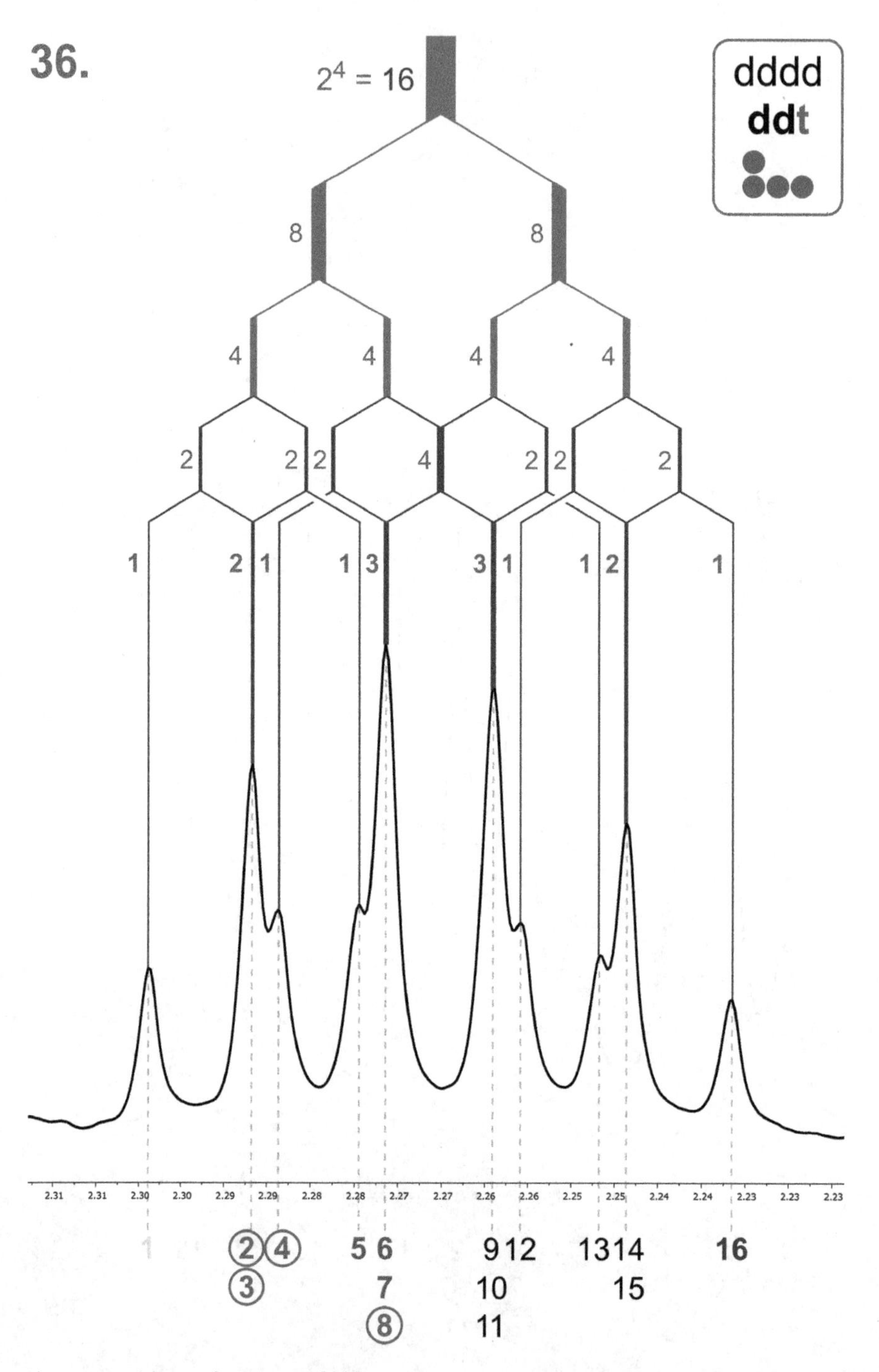

Figure 5.79: Advanced answer 36 [ddt].

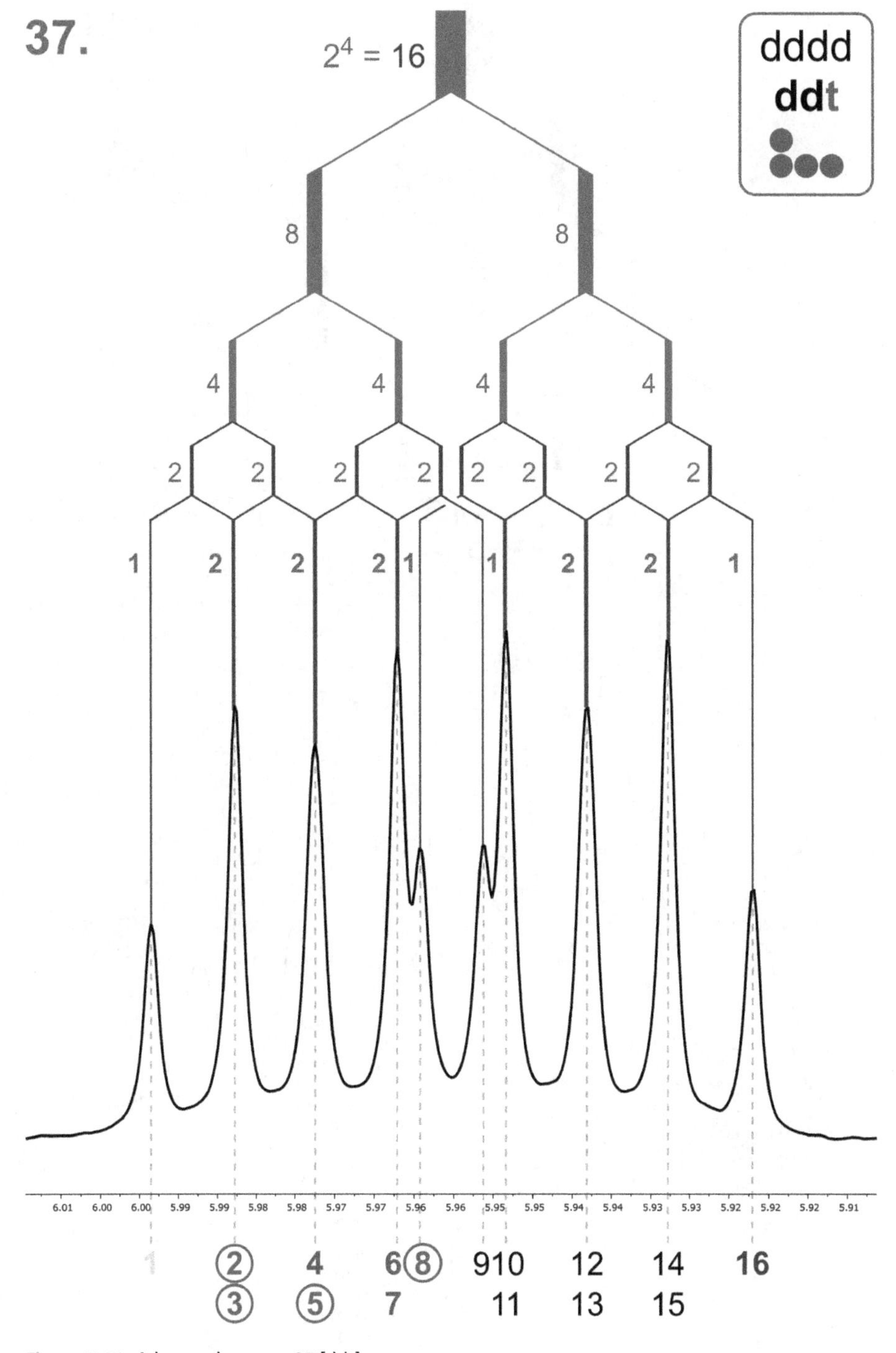

Figure 5.80: Advanced answer 37 [ddt].

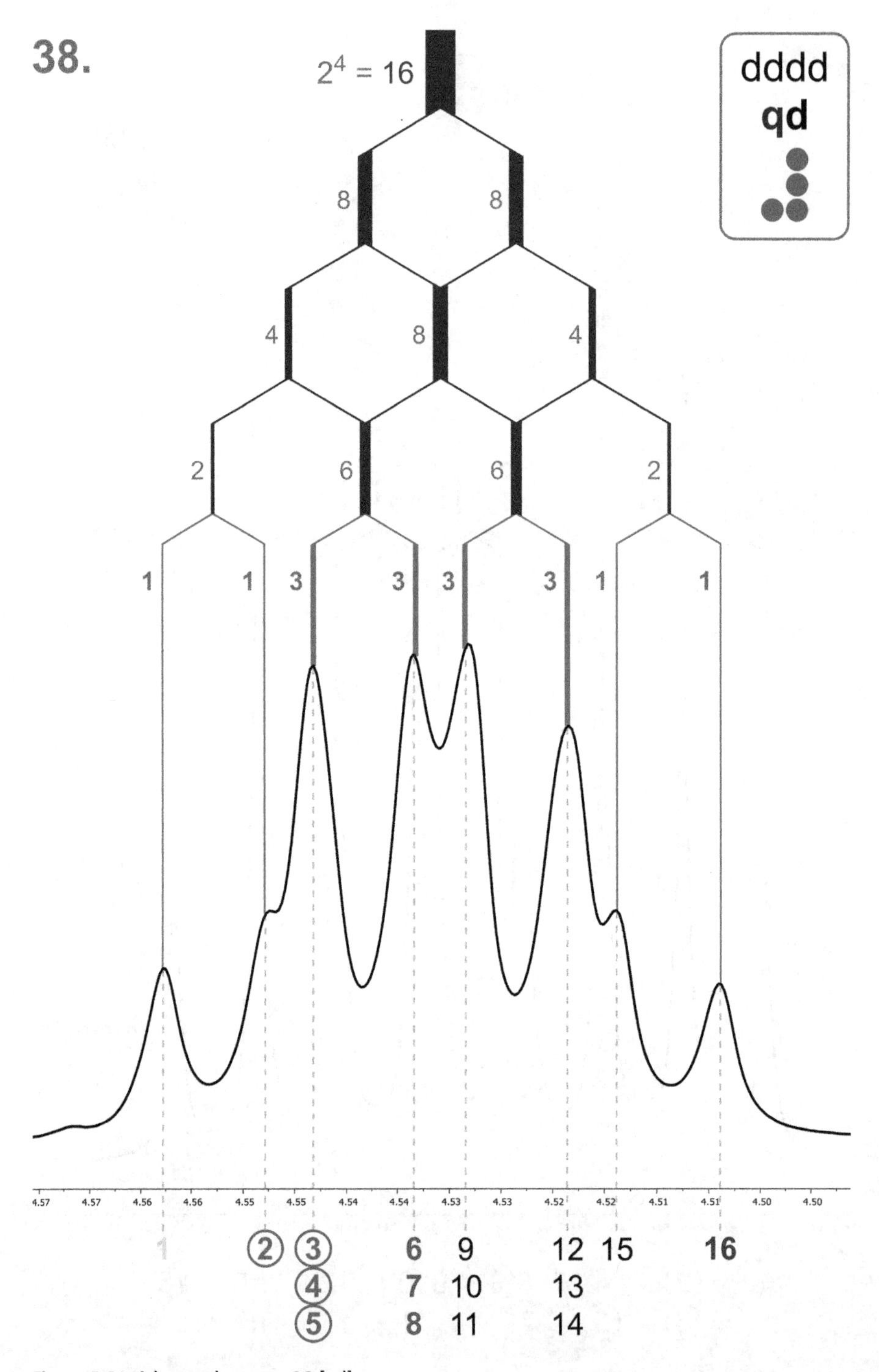

Figure 5.81: Advanced answer 38 [qd].

39.

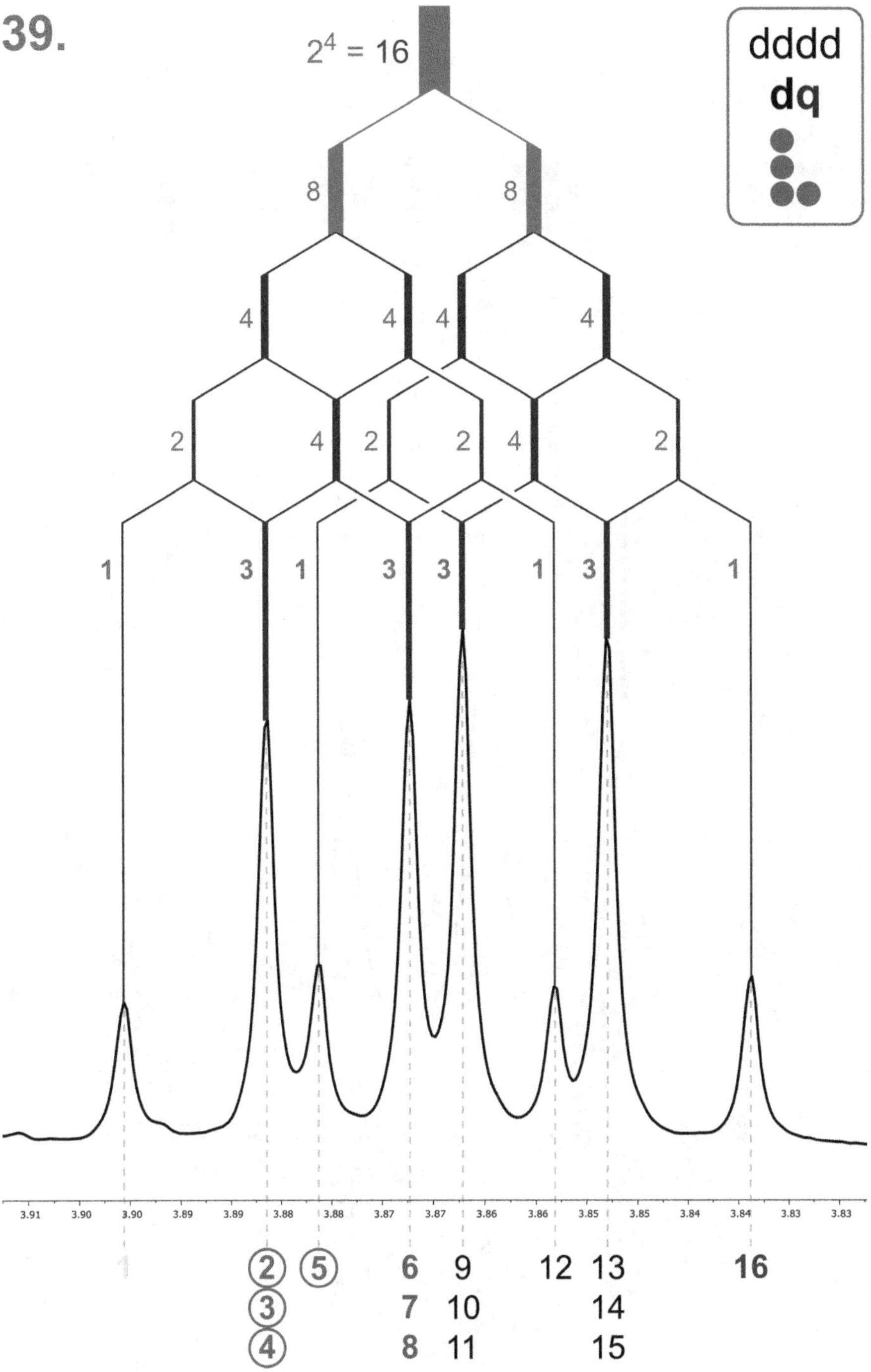

dddd
dq

Figure 5.82: Advanced answer 39 [dq].

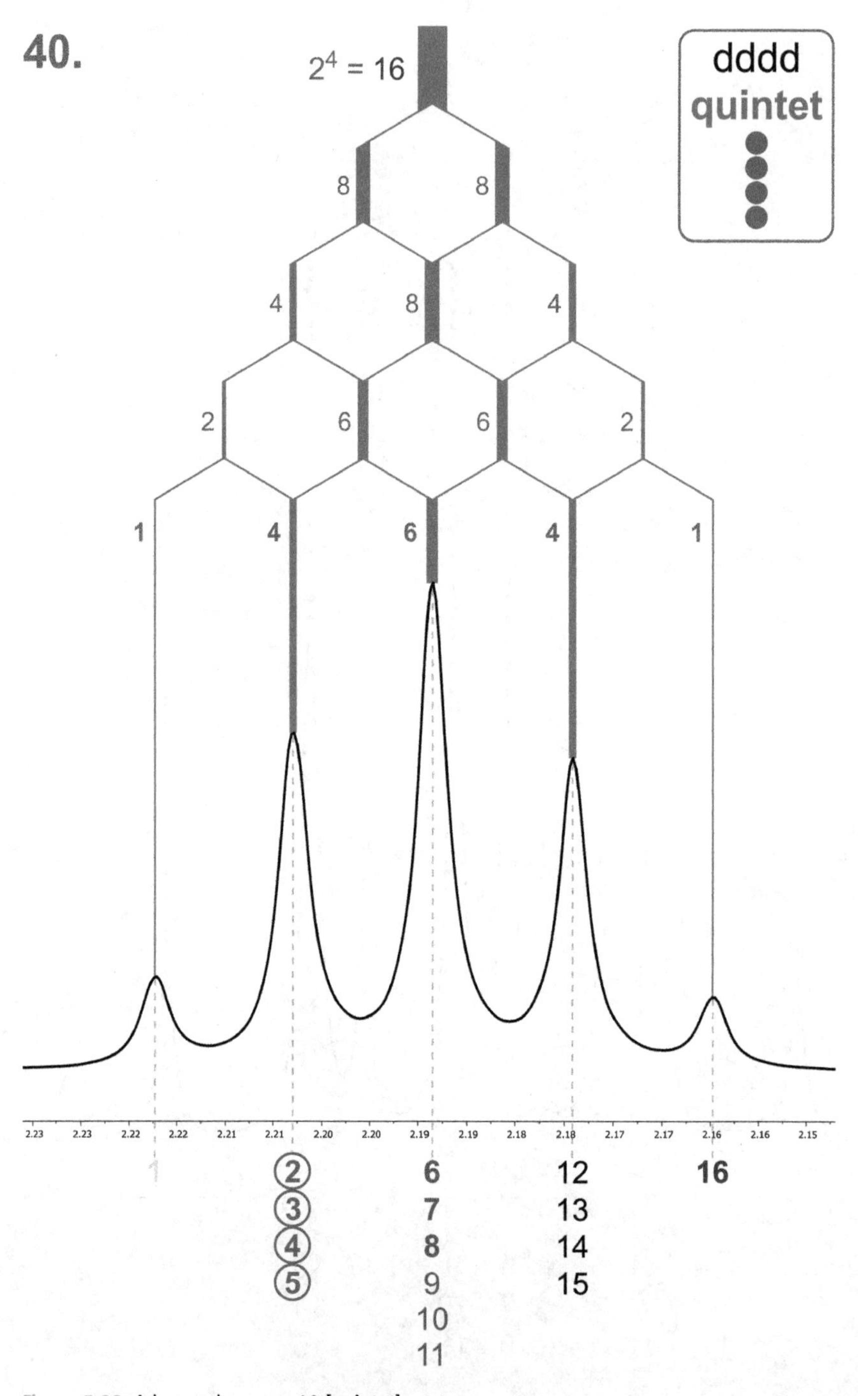

Figure 5.83: Advanced answer 40 [quintet].

5.4 Expert Level: doublet of doublet of doublet of doublet of doublets [ddddd]

$2^0=1$	$2^1=2$	$2^2=4$	$2^3=8$	$2^4=16$	$2^5=32$
0	1	2	3	4	5
s	d	dd	ddd	dddd	ddddd
	doublet	t	td	tdd	tddd
		triplet	dt	dtd	dtdd
			q	ddt	ddtd
			quartet	tt	dddt
				qd	ttd
				dq	tdt
					dtt
				quintet	qdd
					dqd
					ddq
					tq
					qt
					d of quintets
					quintet of d
				sextet	

Figure 5.84: Multiplets with five J-coupling constants ($n = 5$: J_1, J_2, J_3, J_4, J_5).

5.4.1 Expert Exercises

41.

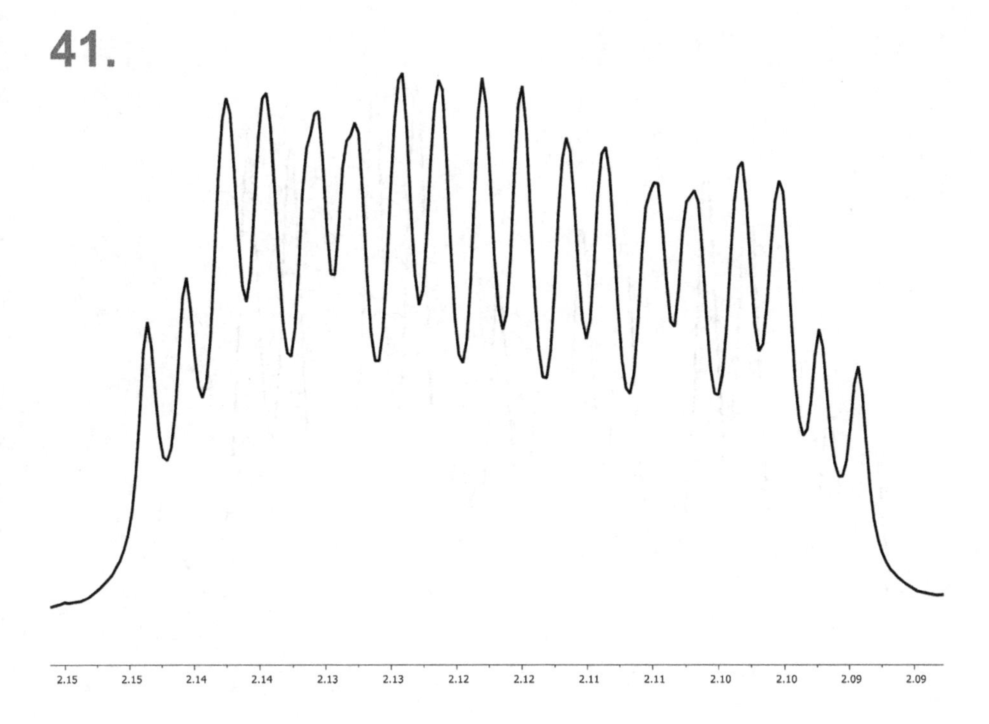

Figure 5.85: Expert exercise 41.

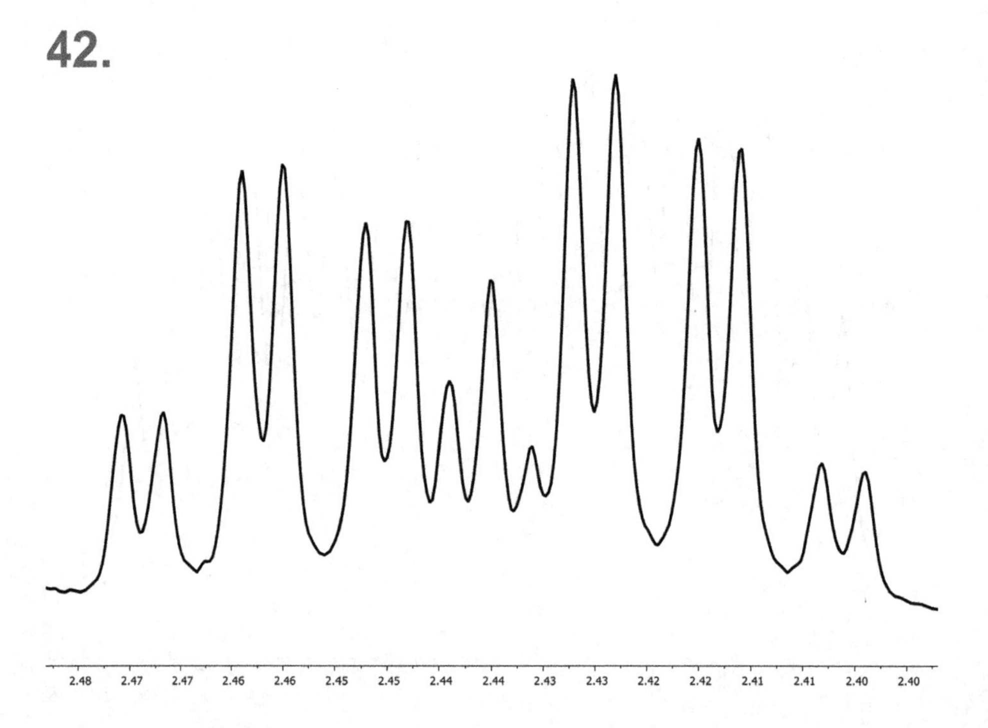

Figure 5.86: Expert exercise 42.

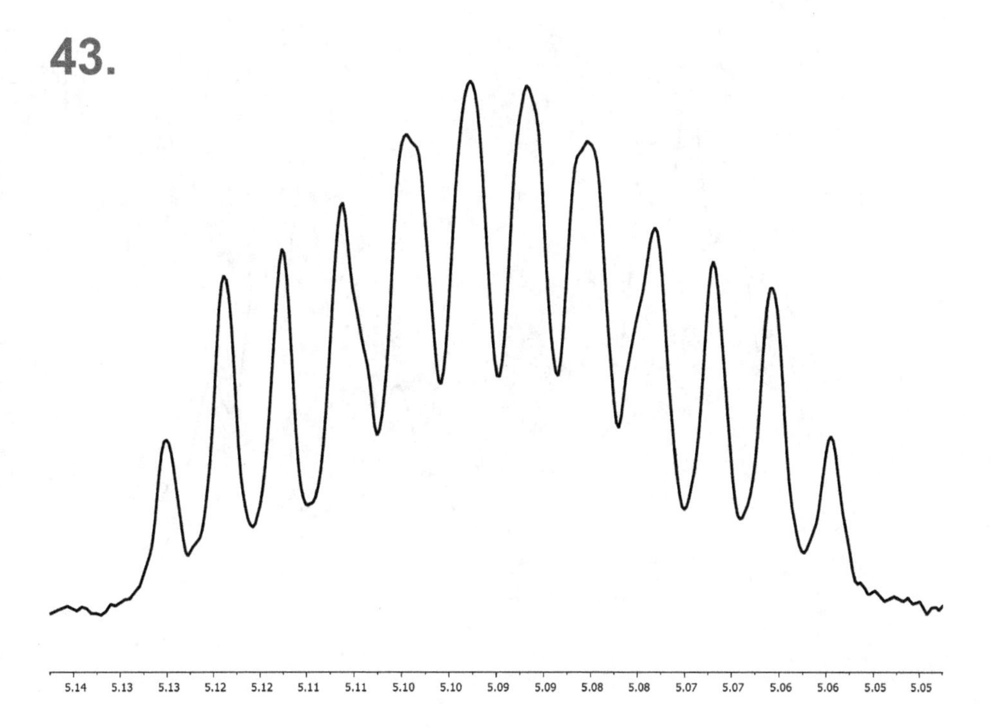

Figure 5.87: Expert exercise 43.

44.

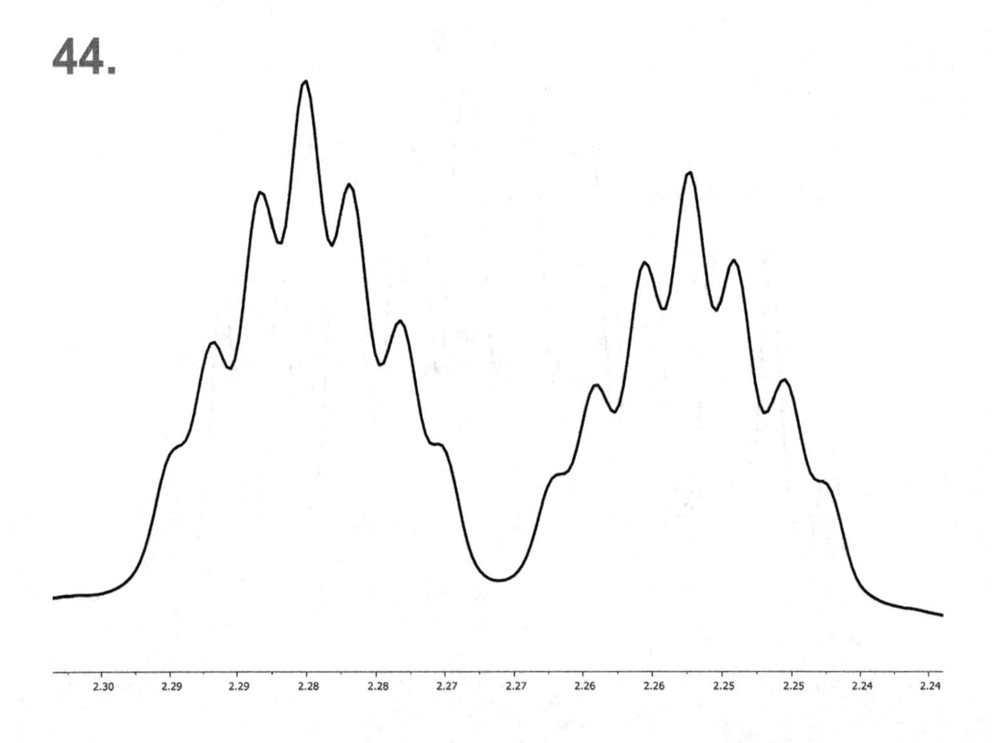

Figure 5.88: Expert exercise 44.

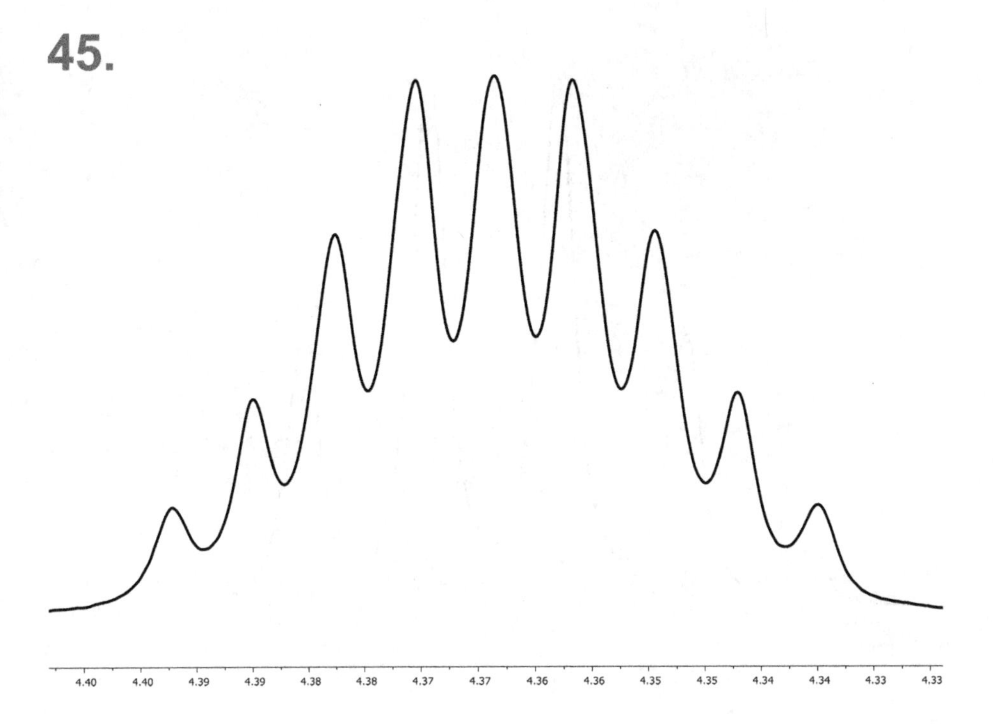

Figure 5.89: Expert exercise 45.

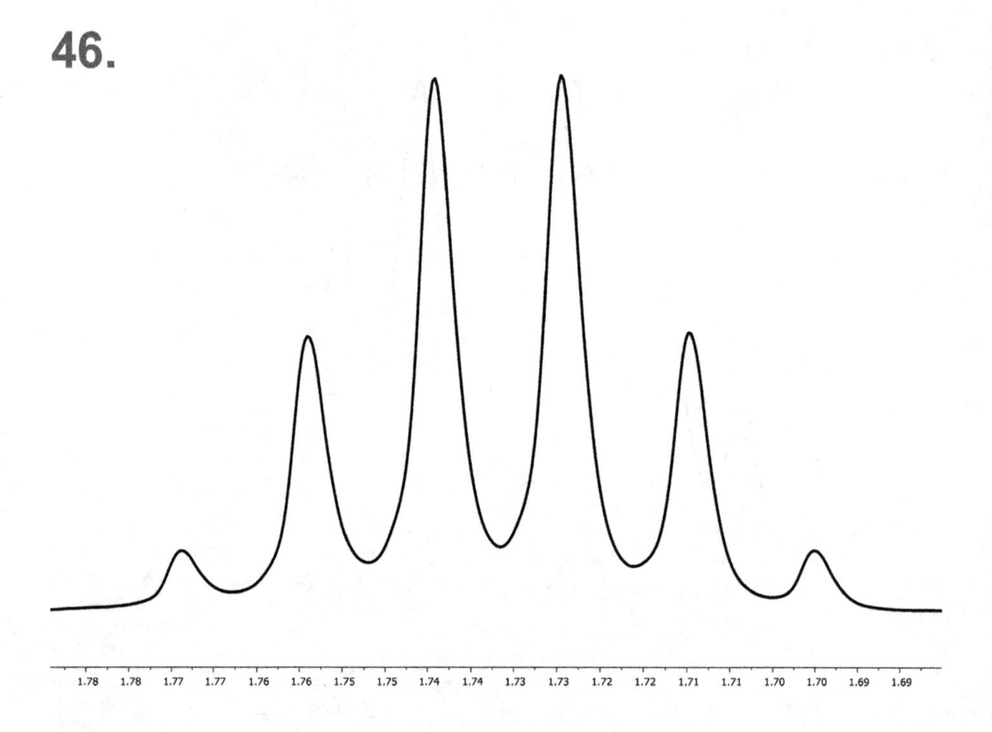

Figure 5.90: Expert exercise 46.

5.4.2 Expert Answers

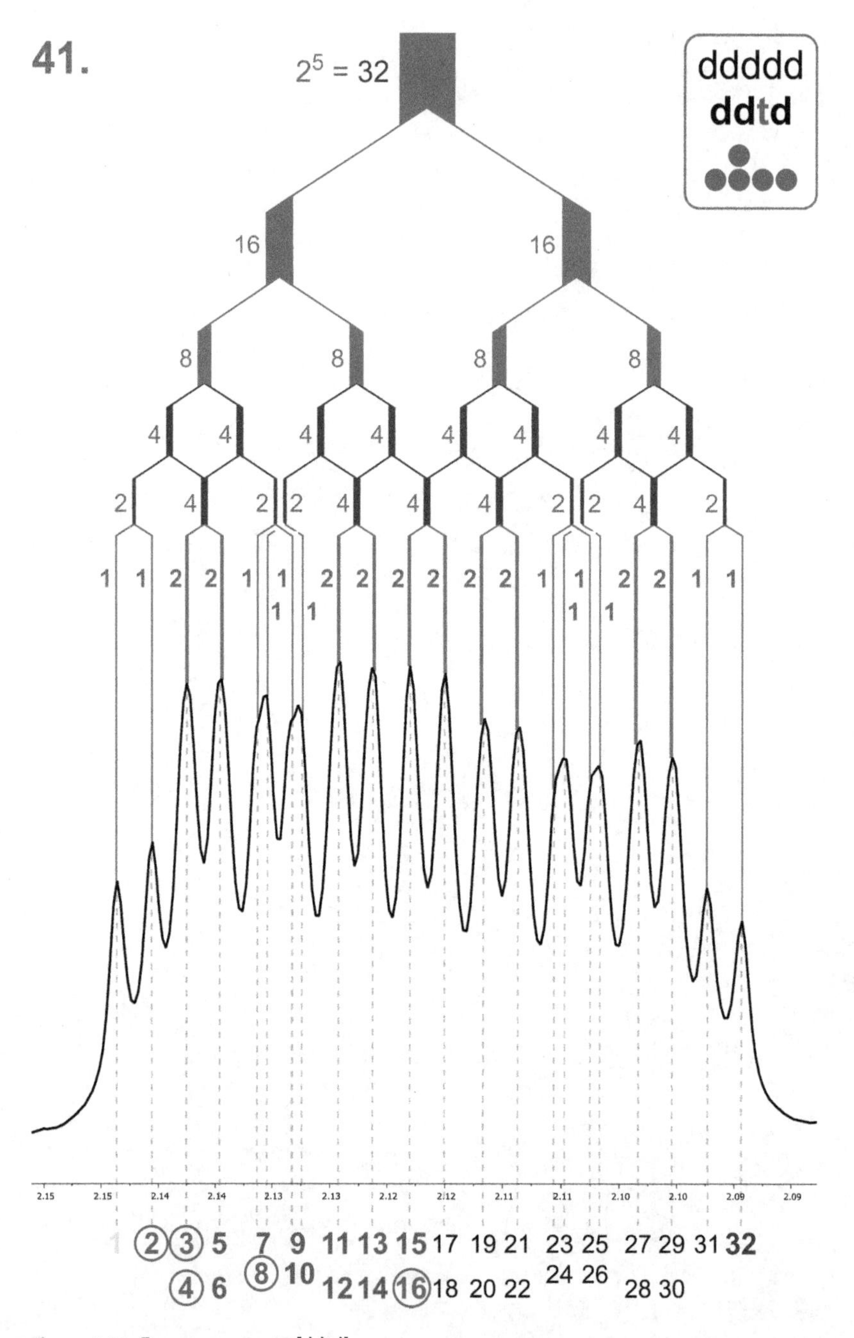

Figure 5.91: Expert answer 41 [ddtd].

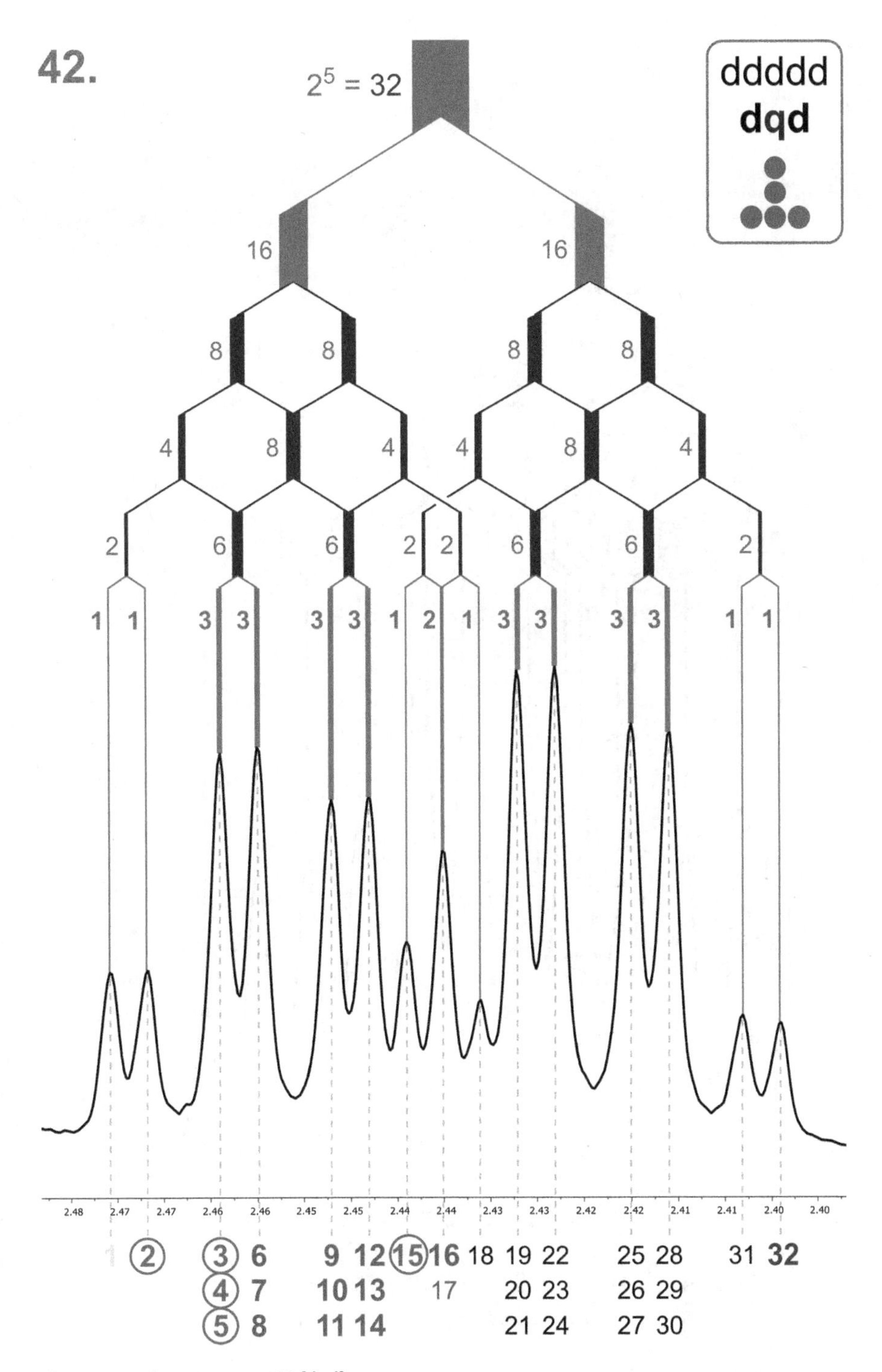

Figure 5.92: Expert answer 42 [dqd].

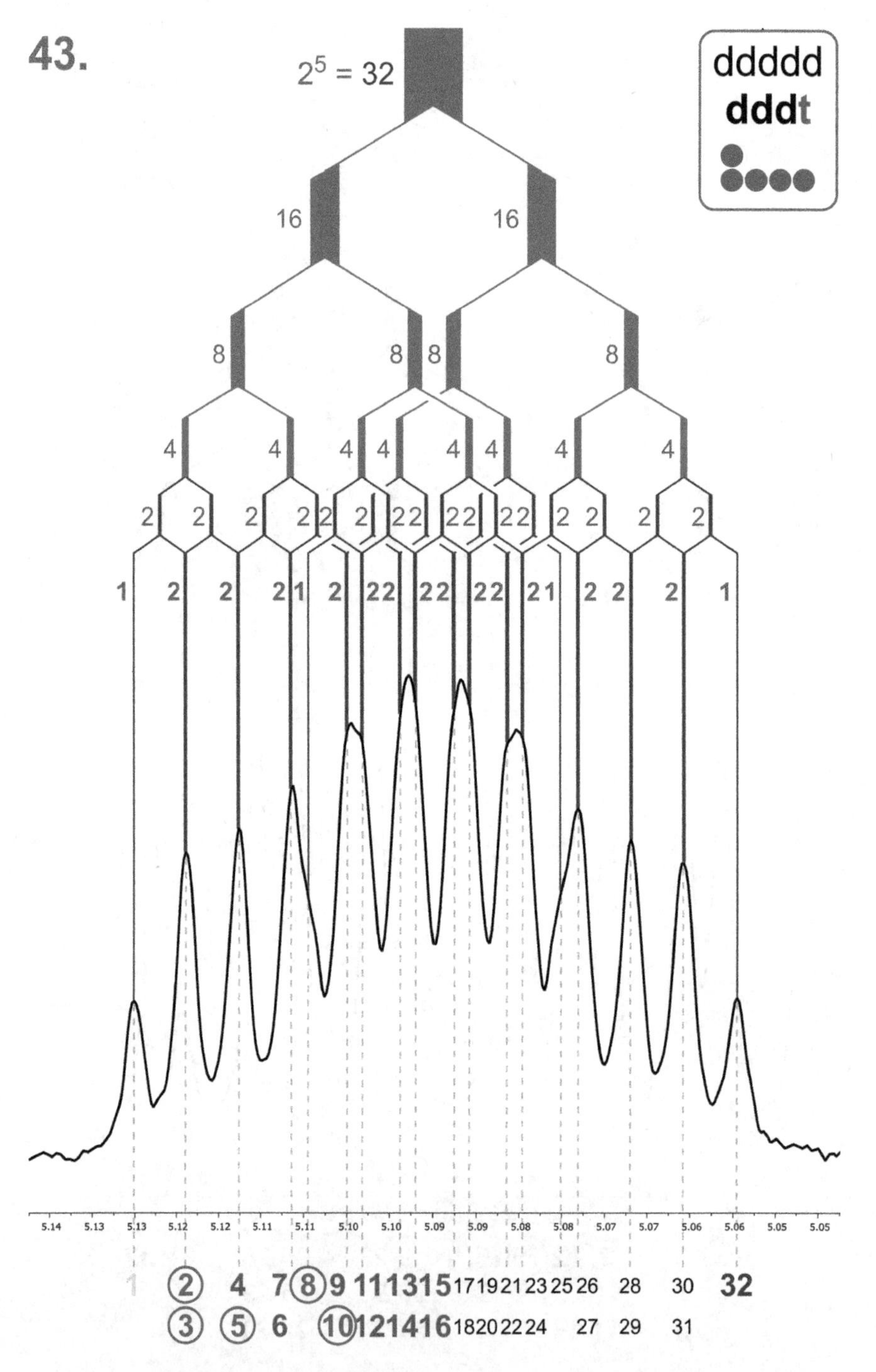

Figure 5.93: Expert answer 43 [dddt].

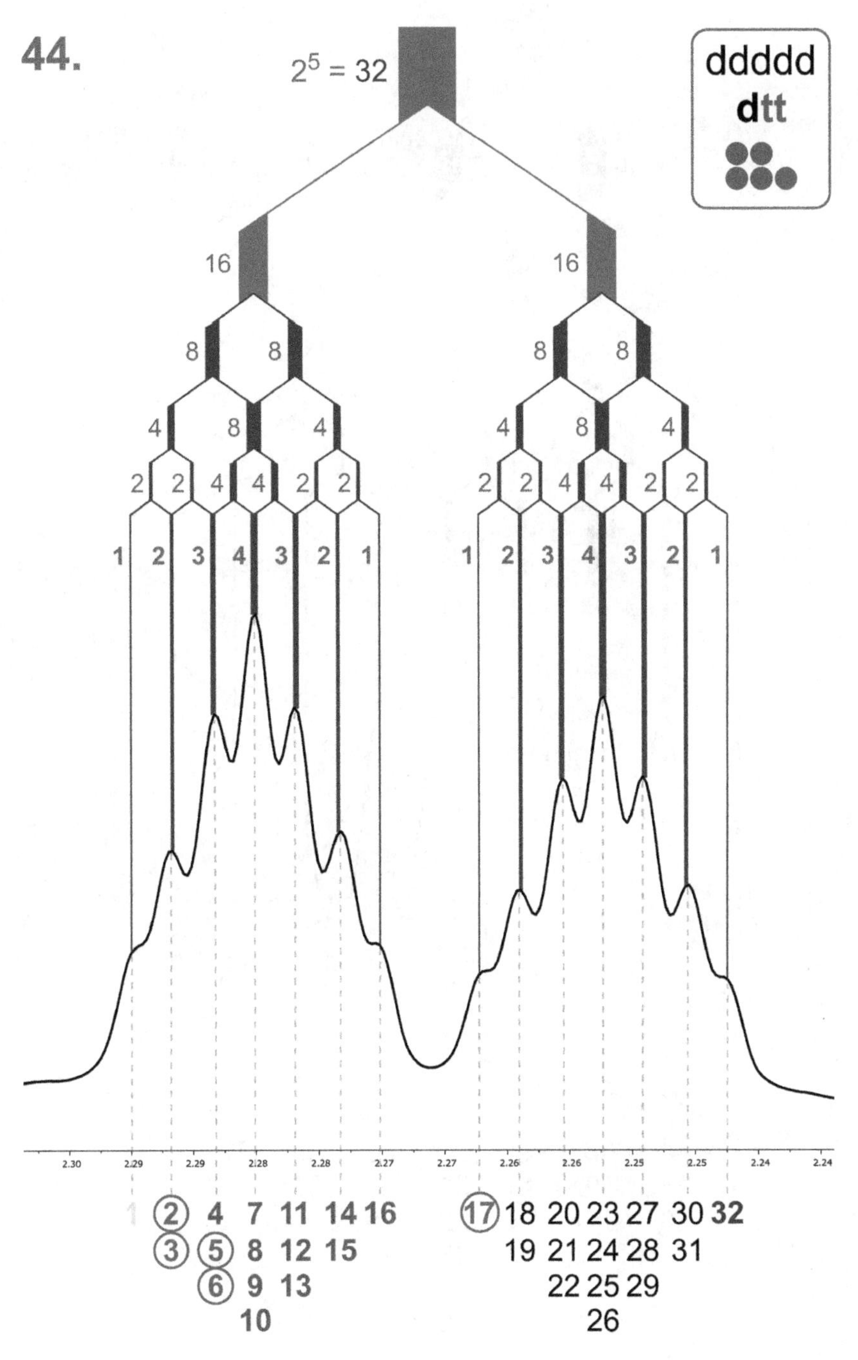

Figure 5.94: Expert answer 44 [dtt].

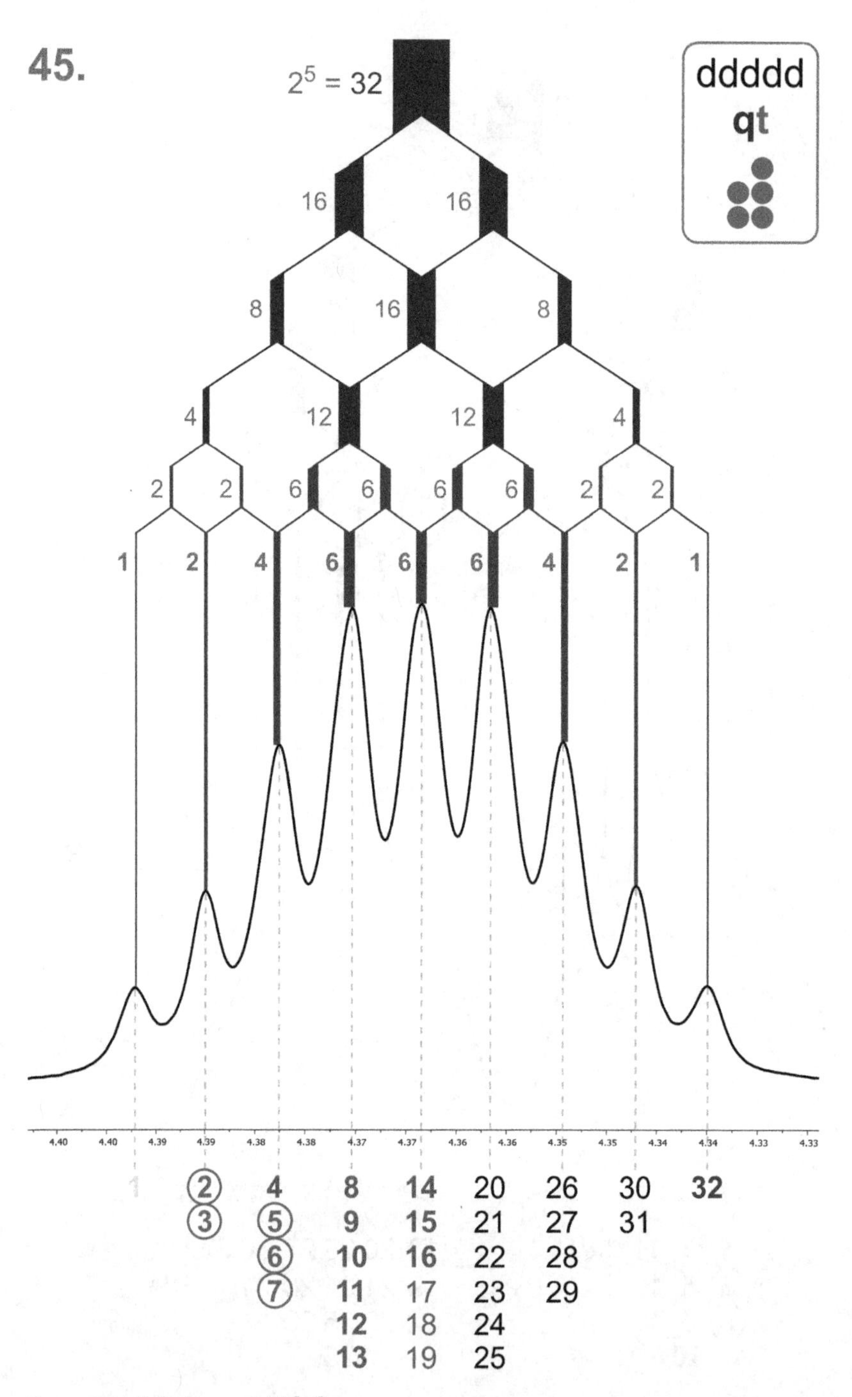

Figure 5.95: Expert answer 45 [qt].

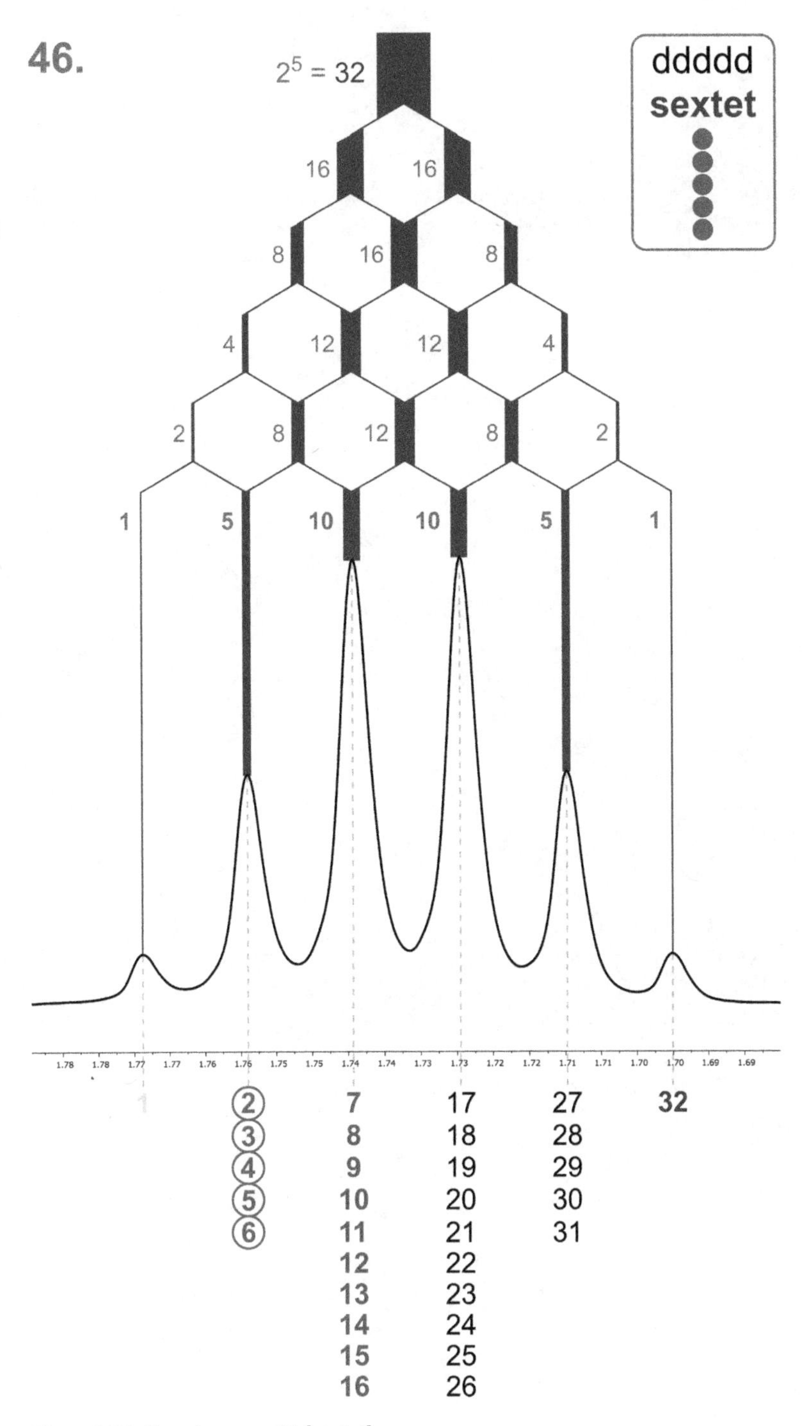

Figure 5.96: Expert answer 46 [sextet].

Appendix

While a graduate student, a postdoctoral research fellow, and a scientist, I always faced a dilemma: how much time I should invest in multiplet interpretation and description? How much value does it bring to the research community, and how accurate are my assumptions? Every scientist should decide for themselves where this golden middle is, without compromising quality, of course. Generally, and for practical purposes, I focus on and report the following signals:

Singlet (**s**)
Doublet (**d**, $J = \ldots$)
Triplet (**t**, $J = \ldots$)
Quartet (**q**, $J = \ldots$)
Doublet of doublets (**dd**, $J_1 = \ldots, J_2 = \ldots$)
Multiplet (**m**) – *for everything else.*

If the multiplet is very characteristic and clear, other complex patterns can be specified, which are depicted in Table 4.1 or Figure 4.3. In some cases, additional details and more-nuanced discussion can be added, for example, specifying that it is a **broad** signal: *broad singlet* (**br s**), *broad doublet* (**br d**), *broad triplet* (**br t**), and so on. However, independent of the level of detail in your description, it will always be a simplification and a compromise between the presentation of the raw data and our interpretation. Only an image of the entire spectrum, or better still the raw data .FID file, will be able to convey full information.

Figures A.1 and A.2 summarize several cases where the appearance of a true **Fundamental Simple First-Order multiplet** (*singlet*, *doublet*, *triplet*, *quartet*) starts to deviate from its ideal shape: cases where it starts to broaden or simply becomes a **Second-Order multiplet.** Please note that it is somewhat controversial to call a second-order multiplet (AA′BB′ or AA′XX′ pattern) an *apparent doublet* (in Figure A.1) or an *apparent triplet* (in Figure A.2). I do not necessarily encourage anyone to follow this nomenclature. However, as an organic chemist, I often use the NMR spectrum of an organic compound as its fingerprint identifier. The appearance of the signal can be more meaningful than its true nature if you need to quickly compare the NMR of the synthesized compound to the reported data or help a peer to reproduce your reported results.

Figures A.3–Figure A.7 are a final collection of colorful infographics summarizing various deuterated NMR solvents in the following format: chemical structure, name, common acronym, and boiling point.[14] You can find additional chemistry infographics at the ChemInfographic Blog.

14 Where the boiling point for the deuterated solvent is not generally known, its nondeuterated analog is reported here in red. All values were summarized from publicly available online sources, for example https://www.sigmaaldrich.com/ (accessed June 15, 2019).

https://doi.org/10.1515/9783110608403-007

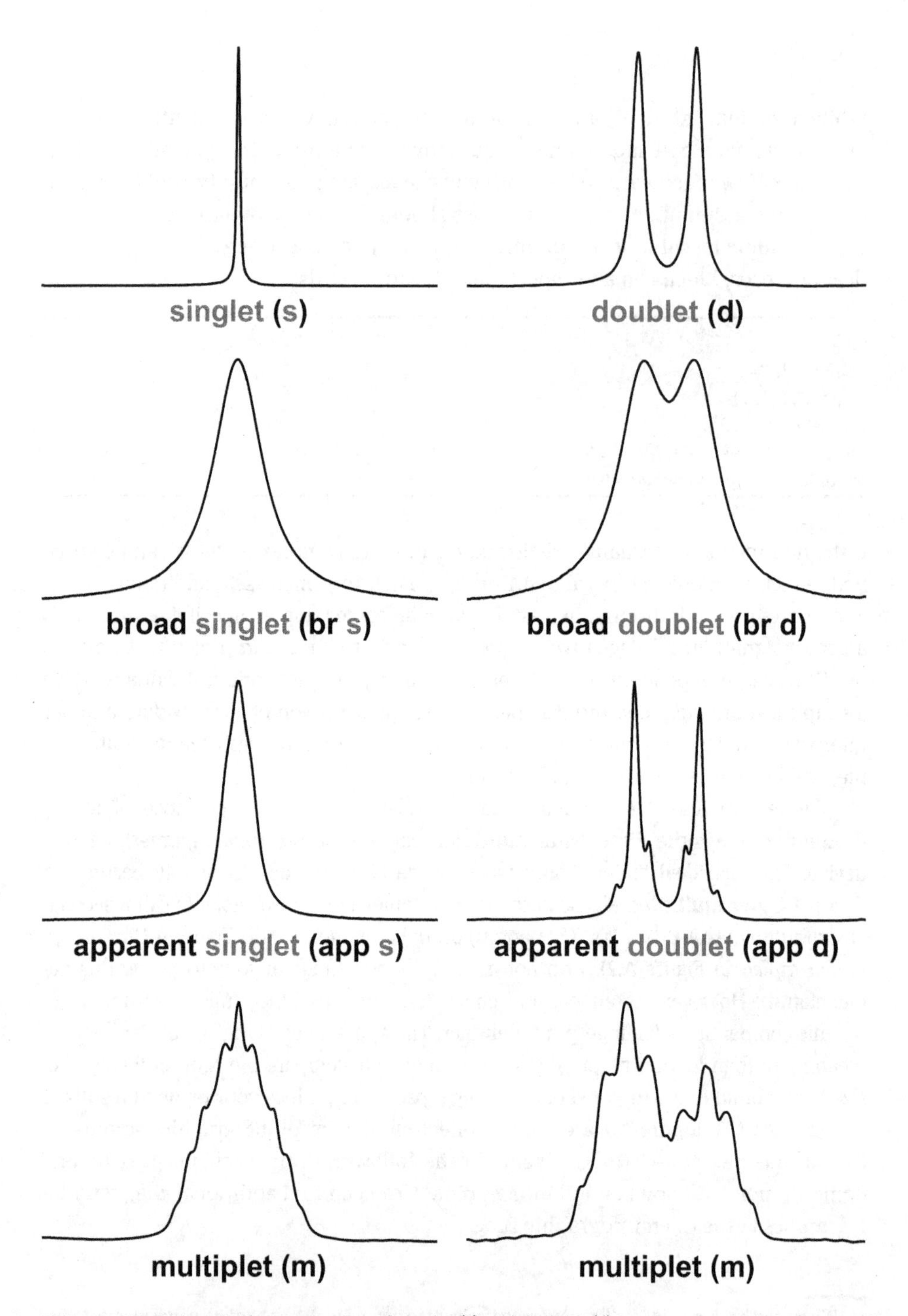

Figure A.1: Variations in the appearance of singlet (s) and doublet (d).

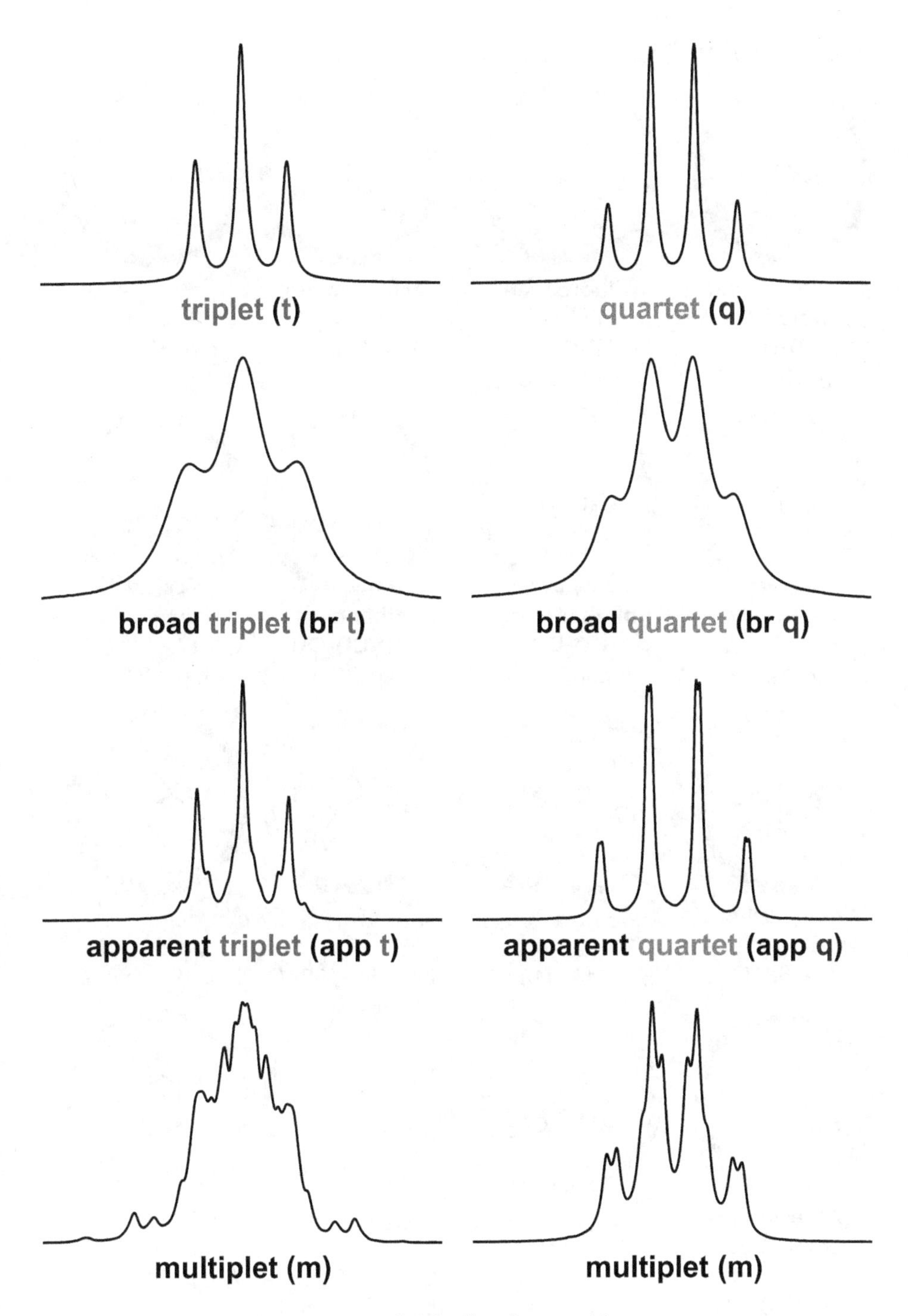

Figure A.2: Variations in the appearance of triplet (**t**) and quartet (**q**).

Protic Polar Deuterated NMR Solvents:

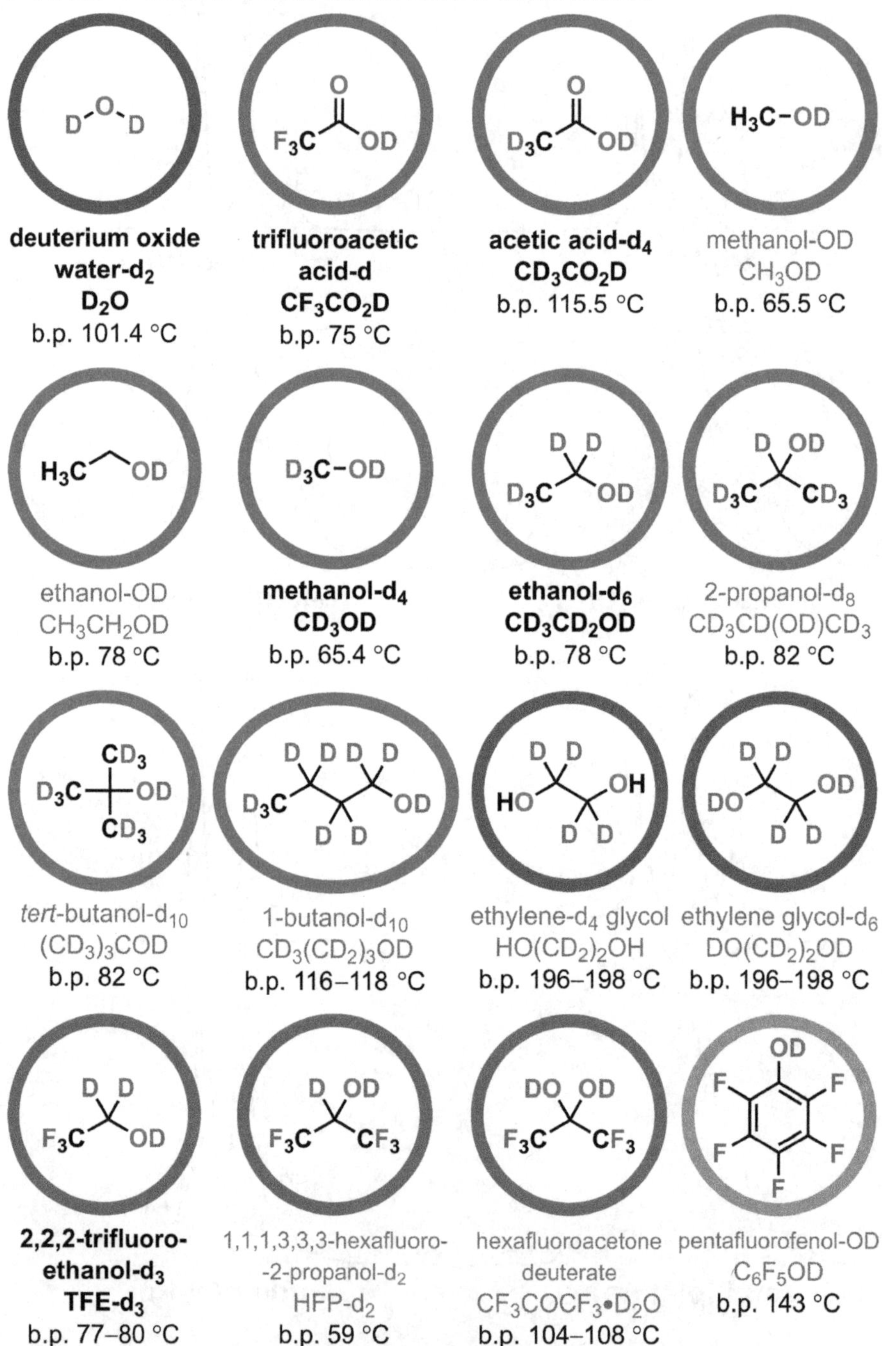

Figure A.3: Protic polar deuterated NMR solvents.

Aprotic Polar Deuterated NMR Solvents:

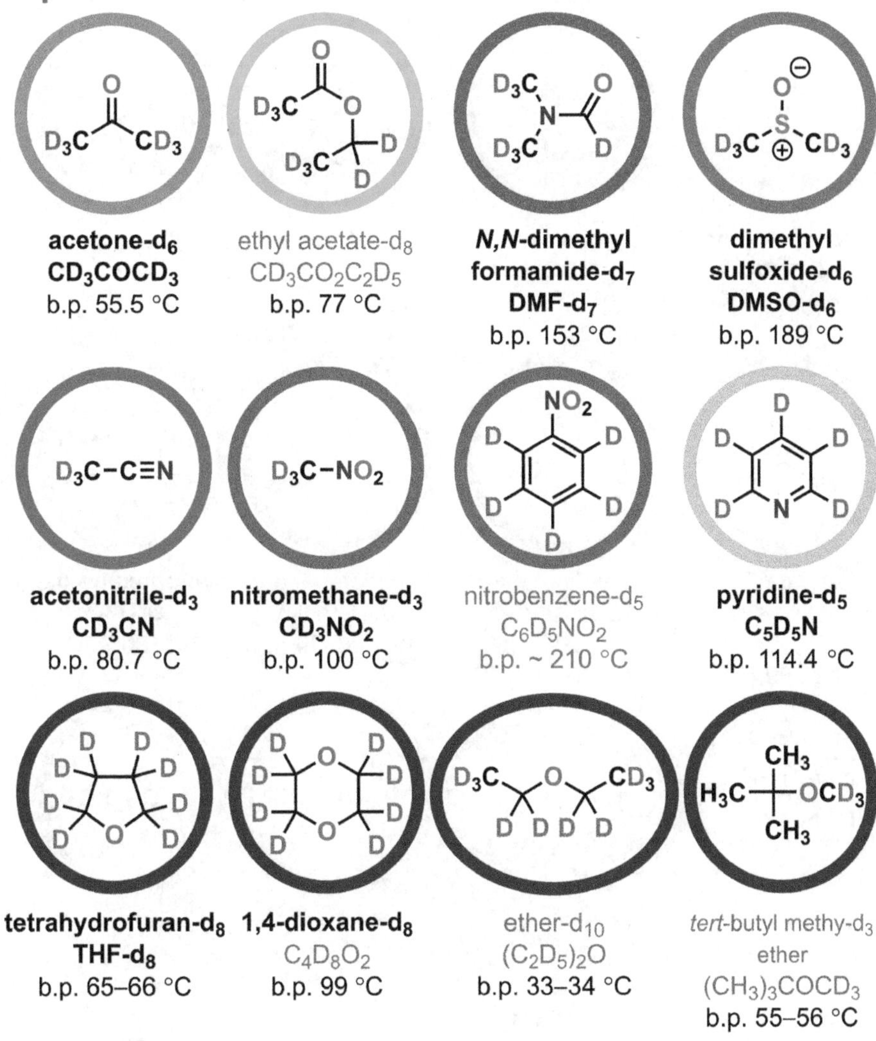

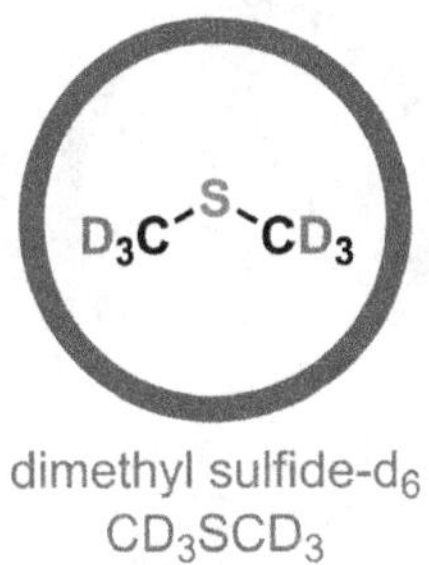

dimethyl sulfide-d_6
CD_3SCD_3
b.p. 36.5 °C

Figure A.4: Aprotic polar deuterated NMR solvents.

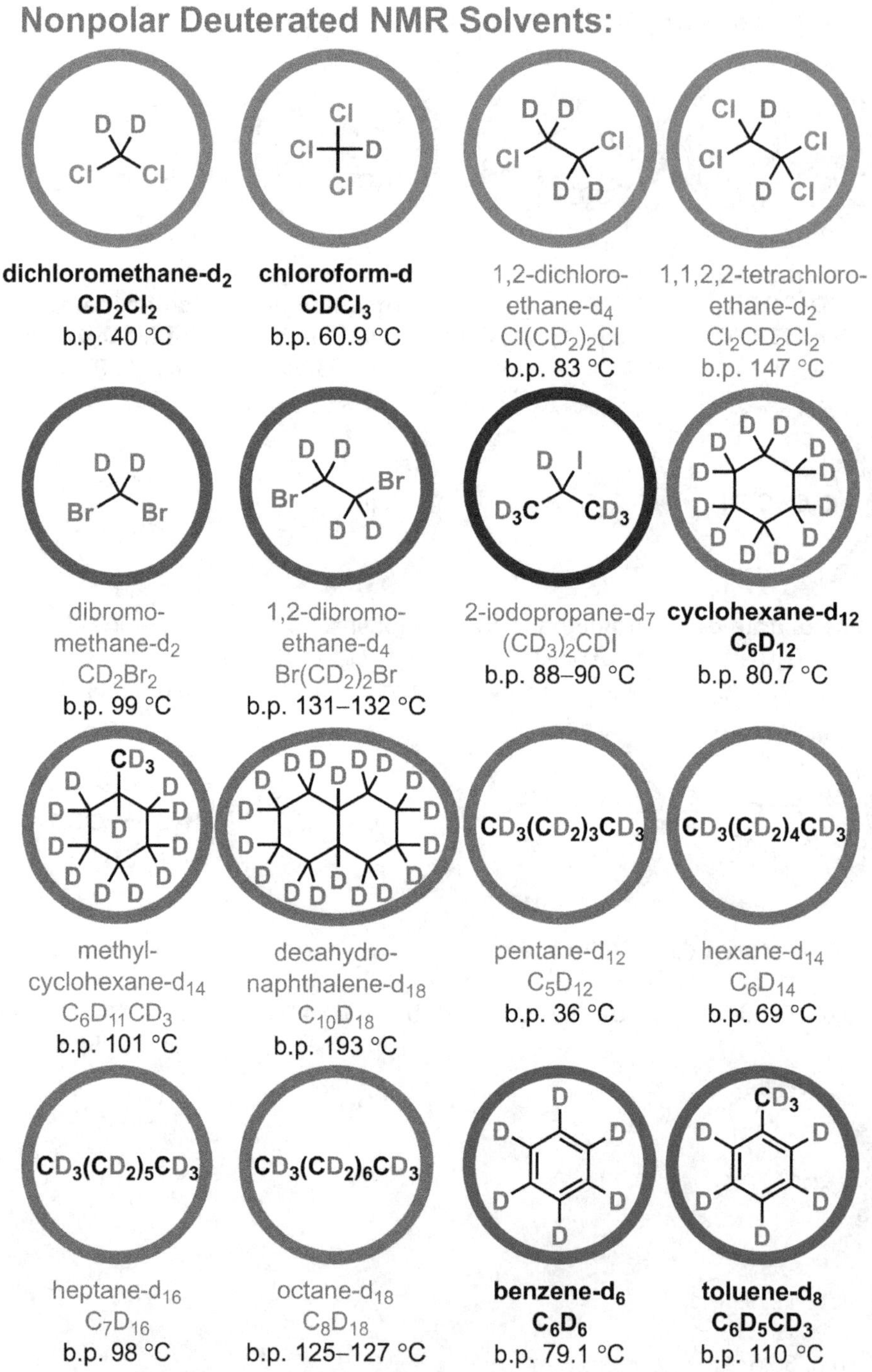
Nonpolar Deuterated NMR Solvents:
dichloromethane-d_2
CD_2Cl_2
b.p. 40 °C
chloroform-d
$CDCl_3$
b.p. 60.9 °C
1,2-dichloro-ethane-d_4
$Cl(CD_2)_2Cl$
b.p. 83 °C
1,1,2,2-tetrachloro-ethane-d_2
$Cl_2CD_2Cl_2$
b.p. 147 °C
dibromo-methane-d_2
CD_2Br_2
b.p. 99 °C
1,2-dibromo-ethane-d_4
$Br(CD_2)_2Br$
b.p. 131–132 °C
2-iodopropane-d_7
$(CD_3)_2CDI$
b.p. 88–90 °C
cyclohexane-d_{12}
C_6D_{12}
b.p. 80.7 °C
methyl-cyclohexane-d_{14}
$C_6D_{11}CD_3$
b.p. 101 °C
decahydro-naphthalene-d_{18}
$C_{10}D_{18}$
b.p. 193 °C
$CD_3(CD_2)_3CD_3$
pentane-d_{12}
C_5D_{12}
b.p. 36 °C
$CD_3(CD_2)_4CD_3$
hexane-d_{14}
C_6D_{14}
b.p. 69 °C
$CD_3(CD_2)_5CD_3$
heptane-d_{16}
C_7D_{16}
b.p. 98 °C
$CD_3(CD_2)_6CD_3$
octane-d_{18}
C_8D_{18}
b.p. 125–127 °C
benzene-d_6
C_6D_6
b.p. 79.1 °C
toluene-d_8
$C_6D_5CD_3$
b.p. 110 °C

Figure A.5: Nonpolar deuterated NMR solvents.

Nonpolar Deuterated NMR Solvents:

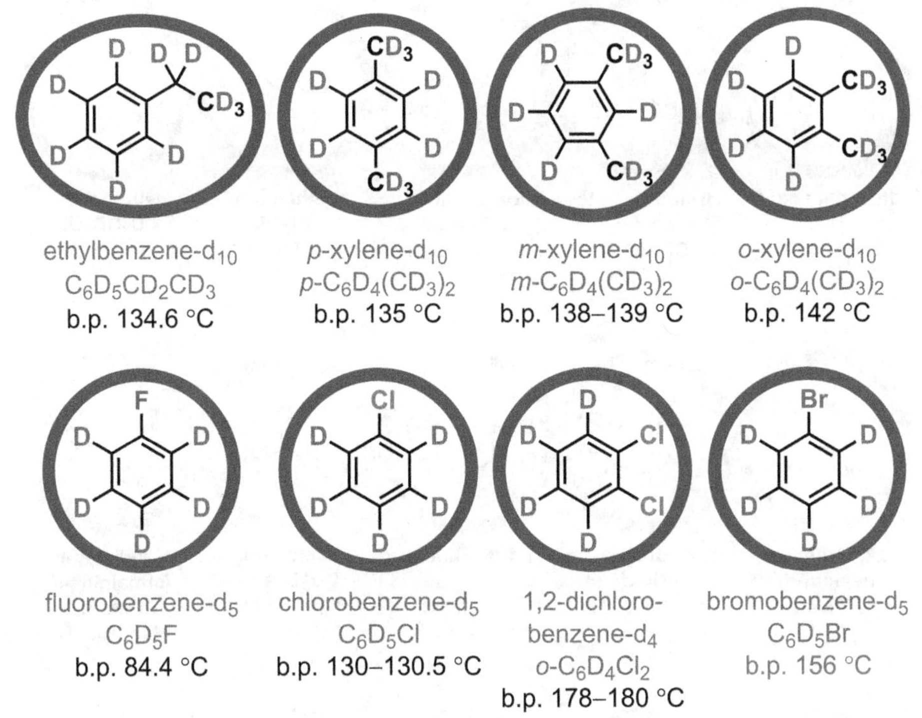

Figure A.6: Nonpolar deuterated NMR solvents.

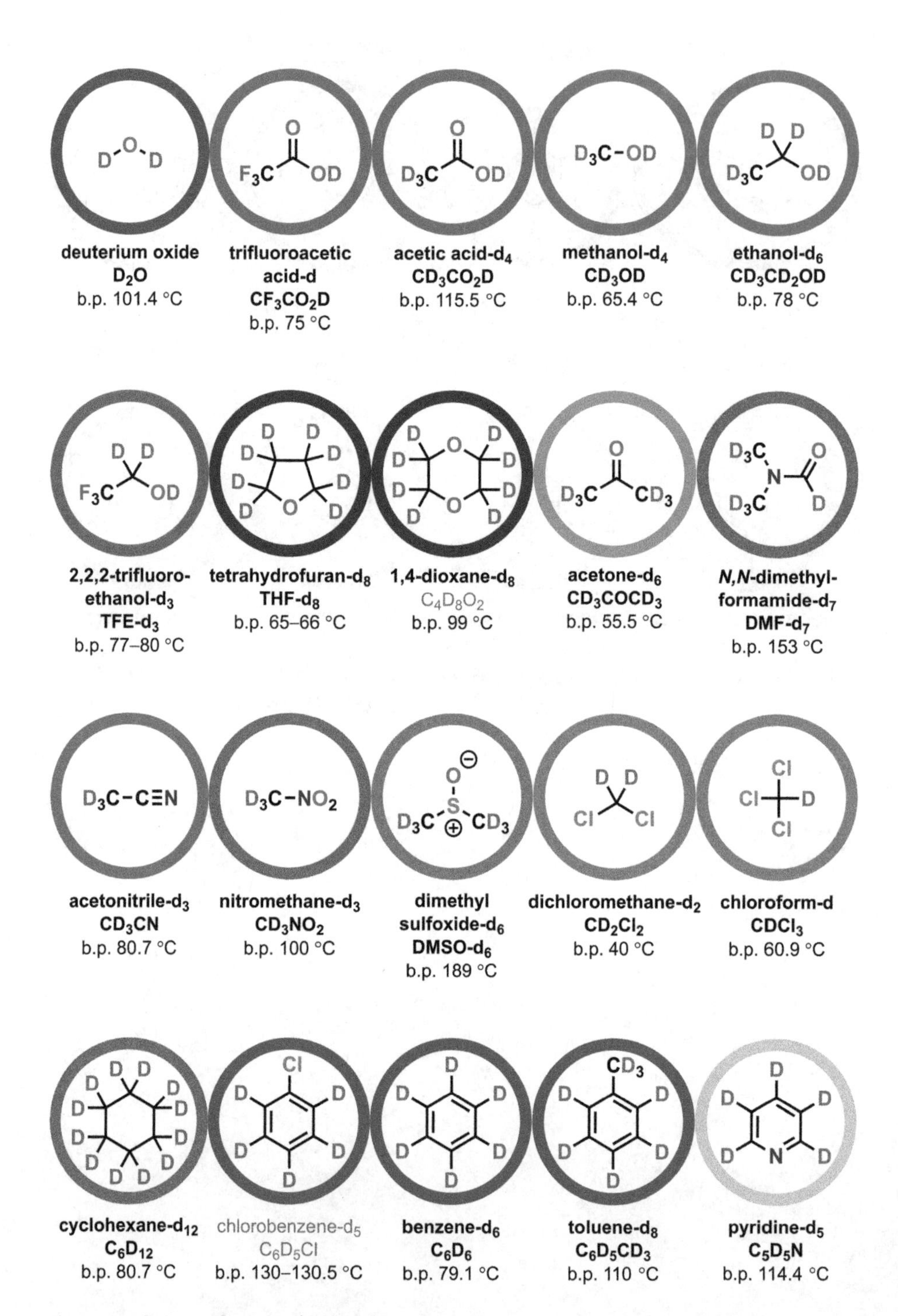

Figure A.7: Common deuterated NMR solvents.

Acknowledgments

I developed all of the visual and infographic content for this book. Any trademarks are of their respective owners and are not intended to suggest or serve as endorsements. The underlying spectra for the Nuclear Magnetic Resonance (NMR) interpretation are drawn from my academic studies and are utilized solely for their shape (the underlying chemical structure is not addressed nor relevant for this workbook). This book does not present original research; however, it does embrace the smart method for deciphering NMR image signals described in a scholarly publication by Professor Hoye's group. Readers are encouraged to reference that original publication for in-depth discussion, technical background, and additional references to previous methods and works. Using the principles from that method, and relying upon my academic, as well as teaching and tutoring experience, and penchant for visual depictions and infographic illustrations, I present a highly visual tutorial for the organic chemistry student and scholar and a set of carefully curated and progressively advanced problems with detailed visual answer keys. The inspiration for this book was partially fortuitous and born out of the strong interest and attention from fellow chemists, scholars, and scientists in the professional community who diligently followed my ChemInfographic Blog and who provided informal, empirical evidence of a need for learning materials that embrace our predisposition for visuals. Special thanks are due to De Gruyter for supporting and embracing a novel workbook that has over 130 figures and visualizations in rich color and for indulging my requirements for a carefully curated presentation of the content with the central goal of optimizing learning.

More fundamentally, the analytical approach via NMR spectroscopy that many organic chemists now take for granted is built on the tireless effort, ingenuity, unwavering curiosity, and ground-breaking discovery of scientists and researchers in the field of NMR. Equipped with that knowledge, we have an invaluable method for elucidating the compounds that we synthesize and develop to advance fundamental research as well as solve problems in society. As a scientist, I am also most grateful for the mentorship, inspiration, and the rich research and learning environments fostered by my graduate and postdoctoral research advisors: Andrei G. Kutateladze, K.C. Nicolaou, and M.G. Finn.

Finally, special thanks are due to my family, in particular, my spouse, friends, peers in the scientific community, and my students. I appreciate the patience and support of my friends and family. And of my students, I am most grateful for the opportunity to teach them, through which I have also further taught myself.

https://doi.org/10.1515/9783110608403-008

References

[1] (a) Kutateladze, AG., & Mukhina, OA. Relativistic Force Field: Parametric Computations of Proton-Proton Coupling Constants in ^{1}H NMR Spectra. J Org Chem 2014, 79, 8397–8406.
(b) Kutateladze, AG., & Mukhina, OA. Minimalist Relativistic Force Field: Prediction of Proton–Proton Coupling Constants in ^{1}H NMR Spectra is Perfected with NBO Hybridization Parameters. J Org Chem 2015, 80, 5218–5225.

[2] Matsumori, N., Kaneno, D., Murata, M., Nakamura, H., & Tachibana, K. Stereochemical Determination of Acyclic Structures Based on Carbon-Proton Spin-Coupling Constants. A Method of Configuration Analysis for Natural Products. J Org Chem 1999, 64, 866–876.

[3] Reynolds, WF., & Enríquez, RG. Choosing the Best Pulse Sequences, Acquisition Parameters, Postacquisition Processing Strategies, and Probes for Natural Product Structure Elucidation by NMR Spectroscopy. J Nat Prod 2002, 65, 221–244.

[4] Kwan, EE., & Huang, SG. Structural Elucidation with NMR Spectroscopy: Practical Strategies for Organic Chemists. Eur J Org Chem 2008, 16, 2671–2688.

[5] Hoye, TR., Hanson, PR., & Vyvyan, JR. A Practical Guide to First-Order Multiplet Analysis in ^{1}H NMR Spectroscopy. J Org Chem 1994, 59, 4096–4103.

[6] Hoye, TR., & Zhao, H. A Method for Easily Determining Coupling Constant Values: An Addendum to “A Practical Guide to First-Order Multiplet Analysis in ^{1}H NMR Spectroscopy”. J Org Chem 2002, 67, 4014–4016.

[7] Silverstein, RM., Webster, FX., & Kiemle, DJ. Spectrometric Identification of Organic Compounds. 7th ed. Hoboken, NJ, USA, John Wiley & Sons, 2005.

[8] Hesse, M., Meier, H., & Zeeh, B. Spectroscopic Methods in Organic Chemistry. 2nd ed. New York, NY, USA, Thieme, 1997.

[9] Günther, H. NMR Spectroscopy. Basic Principles, Concepts, and Applications in Chemistry. 2nd ed. New York, NY, USA, John Wiley & Sons, 2001.

[10] Professor's Hans, J. Reich NMR spectroscopy course is a good example of a comprehensive and heavily illustrated on-line resource. It also has a collection of practice exams and an exhaustive list of NMR bibliography. For more details please visit https://www.chem.wisc.edu/areas/reich/chem605/index.htm (accessed June 15, 2019).

[11] There are numerous free and commercial NMR processing software options: (a) Mnova NMR http://mestrelab.com/software/mnova/nmr/ (b) TopSpin™ https://www.bruker.com/products/mr/nmr/nmr-software/software/topspin/overview.html (c) ACD/NMR https://www.acdlabs.com/products/ and others (accessed June 15, 2019).

[12] Mann, BE. The Analysis of First-Order Coupling Patterns in NMR Spectra. J Chem Educ 1995, 72, 614.

https://doi.org/10.1515/9783110608403-009

CPSIA information can be obtained
at www.ICGtesting.com
Printed in the USA
LVHW060728300321
682888LV00002B/11

9 783110 608359